T0146163

# Disrupting Deterrence

Examining the Effects of Technologies on Strategic Deterrence in the 21st Century

MICHAEL J. MAZARR, ASHLEY L. RHOADES, NATHAN BEAUCHAMP-MUSTAFAGA, ALEXIS A. BLANC, DEREK EATON, KATIE FEISTEL, EDWARD GEIST, TIMOTHY R. HEATH, CHRISTIAN JOHNSON, KRISTA LANGELAND, JASMIN LÉVEILLÉ, DARA MASSICOT, SAMANTHA MCBIRNEY, STEPHANIE PEZARD, CLINT REACH, PADMAJA VEDULA, EMILY YODER

Prepared for the Department of the Air Force
Approved for public release; distribution unlimited

PROJECT AIR FORCE

For more information on this publication, visit **www.rand.org/t/RRA595-1**.

## About RAND

The RAND Corporation is a research organization that develops solutions to public policy challenges to help make communities throughout the world safer and more secure, healthier and more prosperous. RAND is nonprofit, nonpartisan, and committed to the public interest. To learn more about RAND, visit www.rand.org.

## Research Integrity

Our mission to help improve policy and decisionmaking through research and analysis is enabled through our core values of quality and objectivity and our unwavering commitment to the highest level of integrity and ethical behavior. To help ensure our research and analysis are rigorous, objective, and nonpartisan, we subject our research publications to a robust and exacting quality-assurance process; avoid both the appearance and reality of financial and other conflicts of interest through staff training, project screening, and a policy of mandatory disclosure; and pursue transparency in our research engagements through our commitment to the open publication of our research findings and recommendations, disclosure of the source of funding of published research, and policies to ensure intellectual independence. For more information, visit www.rand.org/about/principles.

RAND's publications do not necessarily reflect the opinions of its research clients and sponsors.

Published by the RAND Corporation, Santa Monica, Calif.
© 2022 RAND Corporation
**RAND®** is a registered trademark.

Library of Congress Cataloging-in-Publication Data is available for this publication.
ISBN: 978-1-9774-0933-1

*Cover: Black_Kira/Getty Images.*

## Limited Print and Electronic Distribution Rights

# About This Report

Emerging technologies introduced by competitors will profoundly shape the way that the United States thinks about deterrence, seeks to deter aggression from its adversaries, and provides assurance for its allies. Currently, China, Russia, and other actors are conducting research and development activities in a host of emerging technology areas. Accompanying these activities are potential security implications. Although some research efforts have begun exploring the effect that such technologies could have on operational outcomes, very little is known about their potential relationship to strategic deterrence. Through a fiscal year 2020 study, *Strategic Deterrence in the 21st Century*, RAND Corporation researchers took on the task of identifying the potential connections between emerging technologies and deterrence. This report serves as the study's primary deliverable, summarizing the team's findings on how the interactive practice of deterrence is likely to evolve over the next two decades under the influence of various current and emerging technologies. This research should be of value to the U.S. Air Force and the broader U.S. national security community, because it provides insights into the type of policies and investments that may be required to sustain effective deterrence in the National Security Strategy.

The research reported here was commissioned by the U.S. Air Force's Strategic Deterrence and Nuclear Integration Office (AF/A10) and conducted within the Strategy and Doctrine Program of RAND Project AIR FORCE as part of a fiscal year 2020 project *Strategic Deterrence in the 21st Century*.

## RAND Project AIR FORCE

RAND Project AIR FORCE (PAF), a division of the RAND Corporation, is the Department of the Air Force's (DAF's) federally funded research and development center for studies and analyses, supporting both the United States Air Force and the United States Space Force. PAF provides the DAF with independent analyses of policy alternatives affecting the development, employment, combat readiness, and support of current and future air, space, and cyber forces. Research is conducted in four programs: Strategy and Doctrine; Force Modernization and Employment; Resource Management; and Workforce, Development, and Health. The research reported here was prepared under contract FA7014-16-D-1000.

Additional information about PAF is available on our website: www.rand.org/paf/

This report documents work originally shared with the DAF in June 2020. The draft report, issued on October 1, 2020, was reviewed by formal peer reviewers and DAF subject-matter experts.

## Acknowledgments

We would like to thank Lieutenant General Richard M. Clark and Colonel Dave Rickards, and especially James Brooks, of the Air Force A10 for supporting this research and providing guidance along the way. At RAND, we would like to acknowledge Stacie Pettyjohn, director of the RAND Project AIR FORCE Strategy and Doctrine Program, for her support throughout the study. Special thanks also go to Michael Horowitz and Karl Mueller for reviewing an earlier draft of this report; their insightful comments and feedback strengthened our final product. Finally, we wish to thank Rosa Maria Torres for her work in formatting and preparing this document for publication.

# Summary

## Issue

In this project, we examined the implications of eight specific emerging technologies for both the effectiveness of U.S. deterrent policies and the stability of deterrence relationships.

## Approach

We reviewed U.S. government strategy documents to define the deterrence requirements of U.S. national security strategy. We reviewed existing literature on deterrence, escalation, and strategic stability to develop criteria against which to measure the effects of technologies. We evaluated Chinese and Russian views of the nature of deterrence and their perceptions of emerging technologies. For the technology analysis, we identified eight technology areas for closer examination and conducted in-depth assessments of the current status and emerging potential of each. Finally, we employed four lines of analysis to generate possible causal relationships among the eight technology areas and deterrence outcomes.

## Primary Findings

Our research highlights two overarching conclusions. First, taken as a group, *collections of emerging technologies*—especially in the realms of information aggression and manipulation, automation, hypersonic systems, and unmanned systems—*hold significant implications for both the effectiveness and stability of deterrence*. These risks may call for changes in U.S. policies, operational concepts, and technology development programs. Second, *an emerging transition to new ways of warfare, empowered by these same emerging technologies, poses more-general risks to U.S. deterrent policies than does any single technology*. If the United States is left behind in the technological but also conceptual and doctrinal transition to this new era, both the effectiveness and stability of U.S. deterrent policies are likely to suffer. In addition, our research generated more-specific findings. They include the following:

- *Individual technologies are typically an enabler, not a prime cause, of deterrence failure.* Improved capabilities at the margins are rarely if ever decisive factors in deterrence failure.
- Instead, *the risks of deterrence failure are greatest in scenarios in which multiple technologies work together to exacerbate classic sources of deterrence failure*—which may be precisely the scenario set to emerge over the next two decades.
- *Technology combinations complicate deterrence by offering the potential to hit multiple targets across many attack surfaces simultaneously.* This creates an opportunity for

society-wide paralytic attacks that could undermine deterrence by allowing an aggressor to believe that it could freeze the defender long enough to achieve its desired gains.

- *Technologies have the greatest potential to degrade the effectiveness of deterrence in scenarios involving China.* The cases in which technology poses a more realistic threat to the effectiveness of U.S. deterrent threats are largely limited to two contingencies involving China: Taiwan and the South China Sea.
- *Multiple, interacting forms of automation carry very significant risks, especially for the stability of deterrent relationships.*
- *Many technologies challenge the U.S. ability to deter aggression, coercion, and influence-seeking below the threshold of major war.* Cyber capabilities, disinformation, unmanned systems, biological tools, and even artificial intelligence (AI)–driven decision support systems (DSSs) could strengthen and increase the frequency of bellicose actions in the gray zone.
- *There is a growing potential for information-manipulation technologies, including deepfakes, to contribute to the failure of deterrence.*
- *On the opportunity side of the ledger, the United States could employ emerging technologies to enhance the effectiveness and stability of deterrence in multiple ways.* These include investments in resilience against systemic attack and counter–systems warfare; the use of drone, AI-driven analysis, and cyber capabilities as part of a network of persistent, comprehensive domain awareness and targeting capabilities; networks of new-generation precision-guided weapons married to unmanned aircraft systems (UASs) and DSSs to intensify the threat to any advancing forces; and the transfer of technology, including co-development, to allies and partners to enhance their capabilities to deter and defeat aggression.

## Implications for the U.S. Air Force

Using these findings, we offer specific implications for the Air Force:

- To remain attuned to deterrence risks, focus first on understanding the perceptions of rivals and second on the technology.
- The Air Force should place special emphasis on awareness of both the technology packages in which near-peer adversaries are investing and how they seek to combine them.
- Securing against information network or Chinese "system destruction" attacks is a precondition for effective deterrence and stability.
- The UAS and counter-UAS competition is likely to become a major focus of U.S. defense investments and the stability of deterrent relationships in key theaters.
- Norms, rules, and limits governing technologies could benefit the United States.
- Building relationships with rival air force leaders can provide important benefits.
- Technology integration in support of concepts of warfare will be increasingly crucial.
- The United States will gain significant competitive advantage if it can expand multilateral development of priority systems—including sensing, unmanned aircraft, and precision weapon—for partner or ally self-defense.

# Contents

About This Report ................................................................................................................... iii

Summary ................................................................................................................................. v

Tables ................................................................................................................................. viii

1. Introduction: Purpose of Study ............................................................................................ 1

    Focus of This Report .......................................................................................................... 2

    Organization of the Report ................................................................................................ 3

2. Intersections of Deterrence and Technology: A Framework for Analysis ............................ 6

    Identifying and Prioritizing Technological Areas ............................................................... 6

    Developing a Methodology to Assess the Effects of Technologies on Deterrence ............ 10

3. Principles of Effective and Stable Deterrence .................................................................... 15

    Renewed Importance and Evolving Character of Deterrence ........................................... 15

    Major Goals of U.S. Deterrence Policies .......................................................................... 16

    Criteria for Successful Deterrence: Effectiveness ............................................................. 18

    Criteria for Successful Deterrence: Stability ..................................................................... 20

    Challenges to Successful Deterrence: Narratives of Deterrence Failure ........................... 21

    Competitor Views of Deterrence ...................................................................................... 24

4. Overview of Key Technologies .......................................................................................... 25

    Definitions of Technologies ............................................................................................. 25

    Current Status, Future Trajectory, and Limitations .......................................................... 36

5. Effects of Technologies on Deterrence ............................................................................... 42

    Lens 1: Effects on Deterrence Credibility ........................................................................ 43

    Lens 2: Effects on Deterrence Stability ............................................................................ 48

    Lens 3: Implications of Competitor Views of Deterrence ................................................. 52

    Lens 4: Effects of Technology Combinations on Deterrence ............................................ 54

6. Findings and Conclusions .................................................................................................. 58

    General Findings ............................................................................................................. 59

    Implications for the U.S. Air Force ................................................................................... 62

    Technology, Deterrence, and New Ways of Warfare ......................................................... 64

Abbreviations ...................................................................................................................... 67

References ............................................................................................................................ 68

# Tables

Table 1.1. Technologies Selected for Focus of Our Research..........................................................4

Table 2.1. Potential Technology Areas for Analysis....................................................................8

Table 2.2. Technologies Selected for Focus of Study .................................................................10

Table 2.3. Analytic Lenses for Assessing Effect of Technologies on
Effectiveness and Stability of Deterrence ................................................................12

Table 3.1. Major Objectives of U.S. Deterrence Policies Cited in Strategic Documents .............17

Table 3.2. Key Variables Governing Effectiveness of Deterrent Threats.....................................19

Table 3.3. Criteria for Assessing Effect of Technologies on Deterrence Effectiveness ..............19

Table 3.4. Criteria for Assessing Effect of Technologies on Deterrence Stability .......................21

Table 5.1. Potential Effects on Deterrence Credibility................................................................47

Table 5.2. Potential Effects on Deterrence Stability ..................................................................48

Table 5.3. Implications of Emerging Technologies for Specific Deterrence Challenges .............53

Table 5.4. Potential Effects of Technology Combinations..........................................................56

# 1. Introduction: Purpose of Study

The United States and other leading industrialized nations confront a potential inflection point in social and economic life.[1] A suite of emerging technologies, many of them centered on the processing and manipulation of information and information networks, have the potential to generate revolutionary effects that some have termed a "fourth industrial revolution."[2] The technologies typically highlighted on such lists include integrated information networks (such as the Internet of Things [IoT]), artificial intelligence (AI), autonomous systems (from self-driving cars to unmanned aircraft systems [UASs]), capabilities for manipulation of information and perception, classic cyber infiltration and attack, advanced telecommunication systems (such as fifth-generation [5G] telephony), and emerging quantum computing capabilities. These technologies promise to accelerate the digital and information technology transformation that has been ongoing for four decades and could create both new opportunities and new risks and vulnerabilities, from an insecure IoT to AI-driven automated decisionmaking to highly sophisticated techniques for manipulating human perception.

These technologies will also have significant implications for U.S. defense policy. The RAND Corporation has considered the possible effects of several of them, from AI to information and perception manipulation to autonomous systems to swarming approaches.[3] Some scholars and analysts have argued that the combined effect of many emerging technologies will fundamentally change the character of warfare.[4] Current U.S. defense policy takes this potential seriously in everything from growing investments in AI and autonomous systems to emerging force employment concepts, such as multidomain operations. The precise effect—both individually and collectively—of these technologies remains to be seen, but there is a growing

---

[1] This research was completed in September 2020, before the February 2022 Russian invasion of Ukraine. It has not been subsequently revised.

[2] Klaus Schwab, *The Fourth Industrial Revolution*, New York: Crown Business, 2016.

[3] On AI, see Danielle C. Tarraf, William Shelton, Edward Parker, Brien Alkire, Diana Gehlhaus, Justin Grana, Alexis Levedahl, Jasmin Léveillé, Jared Mondschein, James Ryseff, Ali Wyne, Dan Elinoff, Edward Geist, Benjamin N. Harris, Eric Hui, Cedric Kenney, Sydne Newberry, Chandler Sachs, Peter Schirmer, Danielle Schlang, Victoria Smith, Abbie Tingstad, Padmaja Vedula, and Kristin Warren, *The Department of Defense Posture for Artificial Intelligence: Assessment and Recommendations*, Santa Monica, Calif.: RAND Corporation, RR-4229-OSD, 2019; on information and perception manipulation, see Michael J. Mazarr, Ryan Michael Bauer, Abigail Casey, Sarah Heintz, and Luke J. Matthews, *The Emerging Risk of Virtual Societal Warfare: Social Manipulation in a Changing Information Environment*, Santa Monica, Calif.: RAND Corporation, RR-2714-OSD, 2019; and on swarming approaches, see John Arquilla and David Ronfeldt, *Swarming and the Future of Conflict*, Santa Monica, Calif.: RAND Corporation, DB-311-OSD, 2000.

[4] See, for example, Christian Brose, *The Kill Chain: Defending America in the Future of High-Tech Warfare*, New York: Hachette Books, April 21, 2020; Sean McFate, *The New Rules of War: Victory in the Age of Durable Disorder*, New York: William Morrow, 2018; and Paul Scharre, *Army of None: Autonomous Weapons and the Future of War*, New York: W. W. Norton, 2019.

consensus in both scholarship and policy that such technologies hold the potential to significantly alter the character of war and the threats posed to advanced economies.

## Focus of This Report

In this report, we examine a specific potential implication of emerging technologies for U.S. national security policy: the potential effects of these technologies on deterrence.

Although much of the attention in the national security community so far has focused on the way that these technologies could shape combat itself or change the character of conflict, the initial and, in some ways, primary task for U.S. military power is to prevent war. Much of U.S. global presence and posture aims to deter conflict—in Europe, the Korean Peninsula, the Middle East, and Southeast Asia. New technologies can have different effects on deterrence than they do on military operations. If these emerging fourth-industrial-revolution technologies hold the potential to revolutionize the practice of warfare, they may hold a similar potential to change the practice of deterrence—and create new avenues to deterrence failure. Under the influence of emerging technologies, the character of deterrence could be undergoing a transformation every bit as significant as other changes. But this prospect has not yet been subject to detailed analysis.

Our research sought to address that question through several phases of analysis. We focused on an analytically complex problem: assessing the causal relationship between a loosely bound collection of technologies—whose final form or effect is not yet known—on a geopolitical objective (deterrence) that itself represents a complex mixture of variables. No assessment of such an intricate causal relationship can be definitive. Formal modeling would be of little help because the variables involved are too numerous, ill-defined, and, in some cases, inherently abstract.

Nonetheless, this analysis has produced what we believe are compelling and analytically rigorous findings about ways in which a specific set of emerging technologies is likely to affect the practice of deterrence. The analysis does not allow us to predict *how* likely any of those specific outcomes may be or to claim that they will be the *only* relevant effects. The goal was to identify some analytically grounded risks and opportunities and, in the process, to catalyze thinking about this critical problem.

The project sought to assess the effects of emerging technologies on deterrence over a long-term time horizon—roughly the next 20 years. This is itself analytically challenging because so much about the far end of that time horizon remains unknowable, from the precise shape of emerging technologies by then to the nature of geopolitical relationships. Nonetheless, this analysis self-consciously sought to consider the longer-term, more fundamental implications of emerging technologies on deterrence, well beyond the immediate, short-term effects that may be more straightforward to identify and assess. The 20-year time horizon was, therefore, not so much a precise cut-off point—we cannot quantify where AI or quantum computing or lasers will be at that point—as much as it was a marker of the project's intent to raise long-term and fundamental issues.[5]

Our research examined the effects of technologies on two major aspects of deterrent relationships—their *effectiveness* and *stability*; that is, are deterrent threats credible, do they work, and is the essential equilibrium or stability of a deterrent relationship strong? In terms of *effectiveness*, a technology could undermine the strength of a deterrent threat; if the military applications of AI furnished Russia with an ability to stage a crippling, short-notice attack on the Baltics, U.S. and NATO deterrent threats could be undermined. Because the effectiveness of U.S. deterrence policies is the essential focus of U.S. policy, we considered this to be the initial and primary focus of the analysis—understanding ways in which emerging technologies could undermine the goals of U.S. policy. But technologies can also undermine the *stability* of a strategic relationship, by, for example, massively enhancing the value of striking first on the part of both sides regardless of whether either one's basic deterrence policies are credible.[6] We should identify possible effects of technologies on both measures of the health of a deterrent relationship. The two factors can overlap, and some technologies could affect both: A capability to launch a no-warning first strike would undermine both deterrence effectiveness and stability. But in other cases, a technology might provide a gradually increasing advantage that does not create instability but does call into question the effectiveness of U.S. deterrent threats.

## Organization of the Report

We present the results of our analysis in this relatively short main summary report. The main report describes the phases of analysis in the study, summarized below.

First, we had to develop a specific set of technologies for closer examination from among the dozens that could play a role in shaping the practice of deterrence. This chapter describes the approach we took to generate and evaluate candidates for technologies to highlight in the study. This effort produced a consensus set of eight major technology areas to focus on in our research, listed in Table 1.1.

---

[5] We did, however, seek to identify specific possible or probable developments in each of our technology focus areas across that time frame. In some cases, this allowed forecasts that were more precise than others, but our technology assessments were guided by a 20-year time horizon.

[6] In effect, this distinction reflects the difference between two literatures, each of which we have used extensively in this analysis. One is the basic literature on deterrence—under what conditions do deterrent threats succeed and fail? The other is the literature on strategic stability and such issues as the mutual vulnerability of nuclear forces, which speaks to ways that a deterrent relationship can fail, not because either side comes to disbelieve the other's deterrent threat but because one or both sides believe that the essential strategic relationship is unstable. As we will argue, an important finding of this study is that emerging technologies may have significant effects on both these dynamics.

**Table 1.1. Technologies Selected for Focus of Our Research**

| |
|---|
| Advanced cyber and electronic warfare (EW) to disrupt networked information systems |
| Biotechnology |
| Decision support systems (DSSs) and technologies |
| Directed energy |
| Hypersonic systems |
| Information- and perception-manipulation technologies |
| Quantum information and sensing systems |
| Robotics and semi- and autonomous systems |

We did not identify AI, a broad and complex field involving many different technologies, as a technology area of its own. Obviously, AI is central to any assessment of the future military and social implications of emerging technologies. But it is in their application in specific issue areas that AI and machine learning gain significance. AI as a general field is therefore deeply embedded in our analysis—but in its implications for several of these eight applied technologies.

We then took several complementary steps to assess the problem of deterrence, competitors' views of it, and possible criteria for evaluating the effects of technologies. Chapter 2 summarizes the approach to examining the effects of technology on deterrence. Chapter 3 summarizes the analysis related to the theory and practice of deterrence. We initially sought to define the basic deterrence requirements of U.S. national security strategy by reviewing U.S. government strategy documents. Then, to create a framework for evaluating the implications of technologies for the effectiveness and stability of deterrence, we reviewed existing literatures on deterrence, escalation, and strategic stability to develop specific criteria against which to measure their effects. This work relied heavily on existing RAND research for basic elements of the framework but conducted new analysis of potential avenues to deterrence failure. Finally, because deterrence is an interactive process, it was critical to understand how two primary U.S. competitors—China and Russia—view the nature of deterrence and the possible role of these emerging technologies. We conducted a detailed assessment of each.

In parallel with this review of deterrence literature, we conducted in-depth assessments of each of the eight technology areas. In each case, we asked five major questions:

- How do we define the technology?
- What is the current status of this technology?
- What are its projected near-term military implications?
- What longer-term developments appear likely or possible?
- In what ways could the technology affect deterrence and stability?

Each technology deep-dive examined these five questions about that specific technology area. Chapter 4 summarizes the results of these analyses. The full technology assessments are contained in an unpublished companion volume.[7]

---

[7] Michael Mazarr et al., unpublished RAND Corporation research, 2022.

Finally, we employed four discrete lines of analysis—*lenses*—to generate possible causal relationships between the eight technology areas and deterrence outcomes. Chapter 5 summarizes the findings of these four lines of analyses and the basic study findings on the effects of emerging technologies on deterrence. Chapter 6 offers conclusions and recommendations for the U.S. Air Force (USAF).

# 2. Intersections of Deterrence and Technology: A Framework for Analysis

To assess the effects of technology on deterrence, we needed to navigate two methodological challenges. The first was how to identify the set of technologies to be examined. The second was to determine a rigorous analytical method for assessing the actual effect of these technologies on deterrence. The second challenge goes to the heart of the study's core analytical question: Unlike an assessment of something like nuclear exchange rates or combat modeling, the causal link we examined in this project—the relationship between technologies and deterrence—cannot be derived using quantitative methods.[1] Assessing this causal link ultimately relies on a subjective judgment; we structured the study to make that judgment as rigorous as possible. This chapter briefly summarizes the overall approach used in the study to deal with both these issues.

## Identifying and Prioritizing Technological Areas

This study was not designed to reliably identify the list of technologies with the greatest effect on U.S. strategic deterrence requirements. Nor was it designed to generate a comprehensive list of all technologies that could have such effects. Rather, we aimed to assess the deterrence implications of an initial set of technologies that, although they may have some differing characteristics, represent a set of factors that could have significant effects on deterrence and whose analysis therefore offers insight for the USAF and broader national security community.

The immediate analytical question, therefore, was how to arrive at such a set of technologies for focused analysis. No process could generate an objective hierarchy of such technologies by importance, but we still needed an approach that would allow us to reasonably claim that a given set was somehow representative of the effects of emerging technologies and likely to have a significant effect on deterrence. To do so, we first undertook several lines of research to identify major potential candidates and then conducted an expert elicitation exercise to select the final set of focus technologies.

As a first step, we convened the project team for a brainstorming session on which technologies would be prime candidates for the study. This discussion yielded a preliminary list

---

[1] A good deal of work on deterrence has employed quantitative methods, such as regression analysis, to produce findings. Apart from the reliability of such findings, these methods generally focus on the *historical* effect of a relatively small number of simple variables on deterrence outcomes—such as local military balance or the existence of an alliance. This project sought to assess the *prospective* effects of eight complex technology areas, including their interactive effects, over the next 20 years. This more challenging and forward-looking research question is not subject to the same sort of statistical analysis.

of technologies for consideration, including cyber weapons, space surveillance and warfighting systems, AI (or the collection of tools and capabilities that go by that collective name), hypersonic weapons, 5G networks, quantum computing and cryptography, blockchain technology, swarming unmanned vehicles (air, sea, and land), IoT, and directed-energy weapons (DEWs).

Next, we reviewed official U.S. government critical technology lists to discover which technologies had already been identified as important. Although these lists did not select technologies because of their importance for strategic deterrence, they did nonetheless provide a sense of leading areas of national importance. The lists we examined included the following:

- U.S. Department of Defense's (DoD's) Militarily Critical Technologies List (MCTL) and Developing Science and Technologies List (DSTL)[2]
- Defense Security Service Industrial Base Technology List (which provides an updated version of the MCTL)
- technologies identified in the International Traffic in Arms Regulations (ITAR) export control lists, including the U.S. Munitions List
- Export Administration Regulations export control lists, including the Commerce Control List (600 Series)
- technologies highlighted in the U.S. Department of Commerce's Advanced Notices of Proposed Rulemaking.[3]

One challenge with this line of research is that any review of currently important technologies might miss emerging technologies not yet on peoples' radar. Technologies that could undermine deterrence 25 years from now may exist only in the lab or in theory, and there may be no one conducting any analysis of their military implications today. For example, deepfakes as currently understood did not exist even five years ago and certainly would not have been on any government lists of critical technology. Yet within two years, the technology may well be ubiquitous and could be quite important in campaigns to undermine (or spark) U.S. military action in response to deterrence failure.

Thus, to ensure that we also captured emerging technologies, we conducted a review of analyses and rosters of technologies still in development. These analyses aim to forecast the technology areas likely to be important in another decade or more. To conduct this review, we began by consulting established futures sources, including the World Economic Forum and the National Intelligence Council's Global Trends series, as well as more-specific technology

---

[2] DoD stopped producing or updating the MCTL and DSTL in the early 2000s because of budget cuts, but the MCTL remains a valuable resource because it provided "a detailed compendium of information on technologies which the Department of Defense assesses as critical to maintaining superior U.S. military capabilities" (DoD, *Militarily Critical Technologies*, Washington, D.C., September 19, 2001). Because the MCTL is no longer updated, we did not use it as a major source for currently important technologies but rather reviewed it to see which, if any, technologies consistently appeared on the list throughout the years.

[3] Once the Advance Notice of Proposed Rulemaking takes effect, the list of emerging technologies will become critical technologies under the Committee on Foreign Investment in the United States.

projections by U.S. consulting firms.[4] In addition, we conducted a two-term literature review of "technology" and "deterrence" to gain a general sense of academic work that may have nominated or assessed specific technologies, with a special emphasis on any examples that are still emergent.

Finally, we conducted an initial survey of Chinese and Russian strategic and military literature to identify technologies that the two countries have highlighted as being important to their military doctrine/strategy,[5] as having the ability to undermine U.S. deterrence policies, or as having a significant overall effect on deterrent balances.

These research efforts produced an initial list of 48 potential technology areas for analysis, captured in Table 2.1.

**Table 2.1. Potential Technology Areas for Analysis**

| | |
|---|---|
| 5G networks | Ground systems |
| Additive manufacturing | Hypersonics |
| Advanced computing technology | Lasers |
| Advanced materials | Logistics technology |
| Advanced surveillance technologies | Machine learning technology |
| Aeronautic systems | Marine systems |
| Agricultural | Medical or biomedical |
| Armament and survivability | Microprocessor technology |
| AI | Nanotechnology |
| Autonomy | Nuclear |
| Big data technology | Optics |
| Biotechnology | Position, navigation, and timing technology |
| Blockchain technology | Quantum information and sensing technology |
| Chemical | Radars |
| Cognitive neuroscience/brain-computer interfaces | Robotics |
| Command, control, communication, computers | Sensors |
| Computational modeling of human behavior | Signature control |
| Cryptography | Software |
| Cyber weapons | Space surveillance and warfighting systems |
| Directed energy | Space systems |
| Electronics | Swarming unmanned vehicles (air, sea, and land) |
| Energetic materials | Synthetic biology |
| Energy systems | IoT |
| Geoengineering | Virtual reality |

---

[4] See, for instance, World Economic Forum, *Top 10 Emerging Technologies 2019*, Geneva, Switzerland, June 2019; Christopher A. Bidwell and Bruce MacDonald, *Emerging Disruptive Technologies and Their Potential Threat to Strategic Stability and National Security*, Washington, D.C.: Federation of American Scientists, September 2018; U.S. Department of Homeland Security, and Office of the Director of National Intelligence, *Emerging Technology and National Security*, 2018 Analytic Exchange Program, July 26, 2018; Defense Innovation Unit, *Annual Report 2018*, 2018; National Intelligence Center, *Global Trends: Paradox of Progress*, January 2017; and Scott Buchholz, *Tech Trends 2019: Government and Public Services Perspective*, Deloitte Consulting LLP, 2019.

[5] Sources reviewed include a variety of open-source Chinese- and Russian-language journals and think tank analyses, as well as U.S. government strategic documents (e.g., *National Security Strategy* (NSS), *Indo-Pacific Strategy Report*) evaluating Chinese and Russian threats, capabilities, and interests.

To narrow down this list to a set of eight technologies for deeper analysis, we then conducted an expert elicitation exercise.[6] There was no fully objective, purely data-based way to downselect to a smaller group of technologies for analysis. Because of the nature of these technologies, the larger set is incommensurable in important ways; no single variable or criteria would apply in the same way to all of them, and there was no way to generate measurable rankings that would be truly meaningful. In selecting the smaller set, we sought technologies that were (1) currently related or likely to be related to military and strategic issues, (2) under development or consideration by U.S. competitors, and (3) capable of being developed and applied in the time frame of the study. We felt the best way to identify technologies that best met those criteria was the organized application of subject-matter expertise in those areas.

For this exercise, we distributed the initial list of technologies in Table 2.1 to our entire project team, asking each of them to select the top eight technologies (in rank order) that they viewed as most important to deterrence and to provide their rationale for each selection. To inform their selections, team members were asked to consider whether the technologies met these four criteria:

1. Is there evidence that the technology is likely to be significant during the course of the study's focus period—two decades out—or is there a chance either that it will not have fully matured or that its major effects would be exhausted by then?
2. Is the technology already the subject of extensive technological and strategic evaluation because of its obvious operational significance—i.e., is there already a large literature assessing its military operational effects?
3. Are there readily available countermeasures to blunt the effects that the technology may impose on deterrence?
4. Does the technology carry the potential to enable novel, destabilizing operational concepts for executing, for example, a military *fait accompli* that would undermine U.S. conventional advantages?

Upon collecting the inputs from each team member, we then tallied the votes for each technology to determine the top eight technologies whose connection to deterrence was strongest across the above variables. In the event of a tie between technology areas, we considered the rank order in which people had nominated the technologies as well as the provided rationale for selection. Because some of the nominated technologies were more specific capabilities and others were broader categories of technologies, we further refined the initial list of eight to combine technologies into distinct areas for further study. Ultimately, we arrived at the set of eight technologies listed in Table 2.2 with the potential to transform the national security and deterrence landscape.

---

[6] Our project team consisted of a mix of subject-matter experts on deterrence, Russian and Chinese defense policy and strategy, and a variety of specific scientific and engineering research areas (e.g., biotechnology, machine learning).

**Table 2.2. Technologies Selected for Focus of Study**

Advanced cyber and EW to disrupt networked information systems
Biotechnology
DSSs and technologies
Directed energy
Hypersonic systems
Information- and perception-manipulation technologies
Quantum information and sensing systems
Robotics and semi- and autonomous systems

These technologies represent a range of sometimes divergent characteristics. Biotechnology is a broad area with multiple specific applications, for example; directed energy is a narrower field but still has a number of sub-technologies; and hypersonic weapons is a much more specific and individual area of technology. In part, these differences reflected our judgment of the relative significance for deterrence of specific systems or broader technology areas. In the detailed analysis of the broader technology areas, we did define and consider specific examples to test their potential effect on deterrence. The differences did not pose a barrier to analysis in part because these technologies are not being compared with one another; our methodology did not require that each technology reflect a commensurate variable that could be included in a set to be measured against common criteria. Each was assessed independently against criteria for deterrence success. As just noted, moreover, this list is not meant to be final or comprehensive but to offer a broad-based snapshot of leading emerging technology areas today with a potential to affect U.S. deterrence goals and practices.

## Developing a Methodology to Assess the Effects of Technologies on Deterrence

Having selected technologies to assess, we then had to develop a methodology for determining the effect that they could have on deterrence—both on the effectiveness of U.S. deterrent threats (do they work?) and the stability of deterrent relationships (is there a destabilizing premium on striking first?).

Any effort to evaluate a deterrent relationship confronts the challenge of dozens of variables, many of which will involve factors that cannot be measured, cannot be directly compared, or both. Deterrent relationships are also highly dependent on very idiosyncratic perceptions by individual leaders and leadership groups. Relying on experience with multiple RAND studies on deterrence, we determined that the most appropriate methodology for this problem was the application of a framework of key issues: By surveying variables that contribute to both an effective and a stable deterrent relationship, we could first identify specific requirements for each and then assess the potential effect of technologies on those variables.

To develop the factors for such frameworks, we reviewed two bodies of prior RAND work— on deterrence and on crisis stability or escalation. These again represented the two basic

10

deterrence qualities we were interested in for this analysis: Is deterrence effective—that is, does a deterrent relationship meet the criteria for the threat to work, as discussed in the literature on deterrence? And is the deterrent and broader strategic relationship stable—are there factors that prompt military preemption or other forms of general or crisis instability or that encourage escalatory dynamics during crises or war? The teams behind these RAND studies had conducted extensive literature surveys and qualitative historical case analyses to validate the variables in their frameworks. This work, updated with new and focused analyses on sources of deterrence failure, allowed us to generate sets of historically grounded factors associated with the effectiveness and stability of deterrent relationships. This chapter describes these factors in detail.

Determining the effects of specific technologies on deterrence, then, involved research into the character of each technology and an assessment of the implications it might have for each major variable in the framework. We approached this challenge in three phases.

First, we conducted deep-dive analyses on each of the eight focus technologies listed above.[7] The analyses focused primarily on the nature and potential of the technologies themselves, but each offered initial analyses of the potential effects of the technologies on deterrence broadly speaking, without yet tying those implications to the criteria for deterrence in the two frameworks.

Second, to produce a multiplicity of perspectives and generate the most cross-cutting insights, we formed four distinct teams within the project and assigned each a unique lens through which to assess the effects of technologies on deterrence. Table 2.3 defines these lenses. As it notes, one of the teams explicitly sought to consider the framework for effects on deterrence effectiveness and another considered the effects on stability. The teams' basic task was to use the criteria for deterrence effects outlined in the frameworks and ask how and under what circumstances technologies would trigger them. Two other teams conducted complementary analyses of the Russian and Chinese views of deterrence and their implications for technology effects and of the possible interactive effects of combinations of technologies.[8]

To perform these analyses, each team reviewed the project's assessment of deterrence theory as well as the technology deep dives. The authors of every deep-dive assessment of a technology

---

[7] These deep-dive analyses are in Michael Mazarr et al., unpublished RAND Corporation research, 2022.

[8] That line of research is a particularly important one: In real-world cases of geopolitical tension or competition, one state is unlikely to turn to a single technology to achieve its goal, including a disruption of U.S. strategic deterrence objectives. Rivals will inevitably employ combinations of technologies designed to achieve certain ends. Thus, to understand the implications of a basket of technologies on deterrence, it is critical not merely to assess them in isolation but also to consider various packages that might be employed to achieve effects well beyond the sum of the individual elements. Indeed, the general literature on military success and failure has generally found very few instances of specific, single new technologies having a decisive effect on either geopolitical or military operational outcomes. Typically, countries experiencing large-scale failure in major strategic or military endeavors do so at the hands of adversaries who have knitted together multiple capabilities under a guiding concept of operations that achieves the decisive effect. This, then, is what we should be looking for—the ways in which emerging technologies could empower new or existing concepts of operations that undermine U.S. strategic deterrence objectives.

also prepared a memo for each of the teams summarizing the possible effects of that technology on each specific lens. Using these inputs, each of the four teams produced a write-up of their analysis, which are summarized in Chapter 5 of this report.

**Table 2.3. Analytic Lenses for Assessing Effect of Technologies on Effectiveness and Stability of Deterrence**

| Lens | Examined Variables |
|---|---|
| 1: Affect the power and credibility of deterrent threats | • Affect the strategic balance in ways that produce an urgent concern or imperative in a potential attacker that create a perceived need to go to war.<br>   ○ Affect broader strategic balance and geopolitical standing of major powers, future expectations of the strategic balance by both sides.<br>• Provide decisive and cost-free avenues to aggression to radically shift the cost-benefit calculus regarding war, in general or with regard to a specific target/enemy.<br>   ○ Technologies that offer a particular opportunity for action for which the defender would have no effective and proportional response; could force an adversary to resort to escalatory capabilities.<br>• Offer operational military effects that undermine the credibility of specific deterrent threats.<br>   ○ Have the general ability to impose costs on others or mitigate costs on oneself, including on specific targets isolated by aggressor's operational concepts.<br>   ○ Offer military effects that the defender cannot easily mitigate or counter.<br>   ○ Provide means to paralyze U.S. or allied/coalition power projection in response to an attack.<br>   ○ Provide means to achieve local military goals with great speed, affect the timeline of making deterrent threats under shadow of attack.<br>   ○ Provide means to affect alliance or coalition cohesion.<br>• Offer tools and techniques, including nonmilitary means, for engaging in aggression below the threshold of large-scale force but that nonetheless create new facts on the ground.<br>   ○ Specifically, find ways that create misalignment in responses between the United States and a target of attack.<br>• Affect domestic opinion or shift the balance of power among domestic interests in ways that create a domestic political catalyst for aggression independent of deterrent viability.<br>• Fuel avenues to misperception that undermine the effect of rational deterrent policies. |
| 2: Affect the stability of strategic balances | • Create action-reaction or interactive cycle of both deployments and perceptions that negatively affect foundations of stability.<br>   ○ Distinction between effects of the evolution of technology itself and a U.S. decision to deploy technologies in strategically significant numbers.<br>• Create the potential for successful first-strike options that degrade the inherent second-strike stability of a nuclear or conventional deterrent relationship.<br>   ○ Shift the offense-defense balance toward instability.<br>   ○ Create seemingly cost-free routes to aggression.<br>   ○ Dynamic can also emerge in aggression below the threshold of major conflict.<br>• Carry the potential to destroy the sense of mutual equilibrium and agreement to shared status quo in a strategic relationship.<br>   ○ Create tools that can bring greater effectiveness and power to perceived policy objectives of seeking the overthrow, destruction, or other radical defeat of its rival in ways that do not accept the legitimacy of its regime or rule. |

| Lens | Examined Variables |
|---|---|
| 3: Actor-specific analyses | • Specific U.S. deterrence objectives with regard to Russia and China<br>• Factors affecting Russia's and China's calculus of aggression/action with regard to each U.S. objective<br><br>    o  Effect of technologies on operational concepts of these countries<br>    o  Potential misperception factors fueled by technologies<br>    o  Domestic political situation of each country; degree of risk acceptance of leadership in current strategic context (as well as degree of risk acceptance historically)<br>    o  Constraints on adversaries' ability and willingness to employ these weapons (affordability, level of development for a particular technology, escalation factors)<br><br>• Specific stability perceptions or concerns of Russia and China<br><br>    o  Interaction effects include likely response to technologies by others and effect of action-reaction cycle.<br><br>• Possible effects of technologies on the above factors<br><br>    o  Specific plans by Russia or China to develop/employ these technologies for military effect<br>    o  Russian and Chinese perceptions of the power/effects of these technologies (both in terms of their desire to use them and their fear of adversary use). |
| 4: Combination effects | • Mutual critical dependencies and acceleration/mitigation effects among the technologies include the following:<br><br>    o  Time factors: How do deployment of technologies affect need for and utility of others later on; would deploying one now render another useless later?<br>    o  Specific examples: Hypersonics reduce time for response, thus increasing premium on AI for response control; unmanned aerial vehicles (UAVs) need to communicate; and information- and perception-manipulation technologies can shape the decisionmaking and operational environment and serve as enablers for other technologies.<br><br>• Obvious combination packages in which several technologies, both newly emerging and existing, could have effects on deterrence out of proportion to individual technologies include the following:<br><br>    o  Package of technologies built around cyber/information capabilities as the lead element of an approach designed to cripple U.S. operations<br>    o  A combination of long-range strike systems, including classic missiles as well as hypersonic systems, that could place U.S. deterrent at enough risk when combined with electronic and informational tools. |

Through these lenses, we seek to enumerate the ways in which integrated concepts of operations wielded by competitors and integrating multiple technologies for coherent, mutually reinforcing effects can upset the effectiveness and stability of U.S. deterrence efforts. Combined with the general literature review on technology and deterrence and the initial findings of the technology deep dive chapters, the lenses offer four in-depth analyses of the potential causal relationships between these technologies and deterrence.

For the third and final phase of our research, we conducted a comparative analysis of the conclusions of the four lens teams. We looked for common effects and themes across the lenses and insights related to the ways that the four sets of deterrence outcomes may interact with one another.

In the process of the third phase of our research, we sought to assess the implications of these technologies primarily for what is termed *general* deterrence—the long-term, underlying national policies of deterrence. Our analysis attempted to isolate ways in which these technologies could gradually and in broad terms provide a potential aggressor with greater confidence in their ability to attain goals and/or escape punishment. We did not, however, rule out potential effects on what is sometimes termed *immediate* deterrence—a nation's ability to deter a specific aggressive action in the context of a particular crisis. The analysis therefore applies to both general and immediate deterrence requirements, although with an initial focus on the former.

These three phases yielded a set of carefully structured judgments on the technologies' possible implications for deterrence—judgments that are grounded in awareness of the nature and possible evolution of the technologies and applied to specific criteria of deterrence. We cannot claim that the analysis has identified all potential implications or can rank-order the ones that have been identified. But the methodology brought various established literatures and findings about deterrence together with very detailed considerations of the focus technologies. And, as Chapters 5 and 6 will make clear, several consistent themes did emerge. Even if they are not comprehensive, these should provoke significant discussion because they suggest areas of potential opportunity as well as risk for the United States and the USAF.

# 3. Principles of Effective and Stable Deterrence

To assess the effects of the selected technologies on deterrence, this analysis had to define what is in effect the dependent variable of that relationship—the characteristics of deterrence against which we sought to measure the effects of those technologies. As noted earlier, our assessment of deterrence considered two distinct aspects of its practice: the *effectiveness* and *stability* of deterrent relationships. By *effectiveness*, we mean simply the degree to which a deterrent threat is credible and successful in convincing a potential aggressor not to attack: Does it work? *Stability* refers to the essential equilibrium of a deterrent relationship: Does it suffer from destabilizing pressures to attack first in a crisis?

We sought first to assess the effects of key technologies on U.S. deterrent policies, evaluating whether those policies are credible and to persuade attackers of the likelihood and power of a U.S. response. We also assessed the ways that those technologies affect the stability of deterrence, determining if they create first-use incentives or otherwise generate instability that could upset a deterrent relationship and lead to conflict. Our research sought to identify key criteria for assessing both of those aspects of deterrence.

## Renewed Importance and Evolving Character of Deterrence

*Deterrence*, one recent RAND report concluded, is

> the practice of discouraging or restraining someone—in world politics, usually a nation-state—from taking unwanted actions, such as an armed attack. It involves an effort to stop or prevent an action, as opposed to the closely related but distinct concept of "compellence," which is an effort to force an actor to do something.[1]

The United States, for instance, seeks to deter Russia from invading the Baltics, or North Korea from invading South Korea. It does so with threats of both *denial* (using military force to keep the aggression from achieving its local objectives) and *punishment* (striking back in various ways to impose costs on an aggressor).

For nearly three decades—from the end of the Cold War through the post-9/11 era's focus on counterterrorism—the theory and practice of deterrence lost prominence in U.S. national security debates. Simply put, unlike in the Cold War environment, the United States no longer had major challengers who posed such serious threats that deterrence needed to rank among U.S. national security priorities. Russia seemingly no longer posed a risk of conventional aggression in Europe, and nuclear deterrent requirements became a background condition rather than an immediate requirement. The few threats that continued to demand a deterrent role, such as North

---

[1] Michael J. Mazarr, *Understanding Deterrence*, Santa Monica, Calif.: RAND Corporation, PE-295-RC, 2018, p. 2.

Korean aggression against South Korea, were manageable enough that deterring them became in effect a lesser task for U.S. national security policy—that is, the national security capabilities that the United States developed for its general defense requirements were potent and credible enough to satisfy regional deterrent objectives without special emphasis.

With the rise of a new era of strategic competition against near-peer rivals, however, deterrence has once again become a priority. The United States has renewed concern about possible threats to its deterrent policies in Europe, the Indo-Pacific, and the Middle East. The urgency of getting deterrence right—and possible risks to the practice of deterrence—have not been so great for decades.

At the same time, the emergence of new technologies, many focused on the information systems central to operating modern economies and militaries, is arguably transforming the character of deterrence. The world may be on the verge of an inflection point, moving into what has been called the fourth industrial revolution, wherein a host of interrelated technologies— networks of smart devices, AI, advanced manufacturing techniques, quantum computing, DEWs and many more—will "fundamentally alter the way we live, work, and relate to one another"[2] and, by extension, change the character of warfare.[3]

These two trends—the renewed importance of deterrence and the rise of technologies that may shape its character—represent the basic motivation for this study. As the practice of deterrence grows increasingly important for the United States, our research sought to identify ways in which the requirements for successful deterrence and threats to its effectiveness could change under the influence of emerging technologies.

## Major Goals of U.S. Deterrence Policies

To gain a better understanding of the deterrence requirements that the technologies could affect, we surveyed official U.S. government national security documents—including the National Defense Strategy, National Military Strategy, NSS, National Cyber Strategy, and National Security Space Strategy—to identify the spectrum of adversary behavior that the United States seeks to deter. Table 3.1 summarizes these objectives.

---

[2] Schwab, 2016.

[3] See Table 4.1 and Chapters 4 and 5 for specific examples of how advances in technology could have transformative effects on the conduct and character of warfare.

**Table 3.1. Major Objectives of U.S. Deterrence Policies Cited in Strategic Documents**

| Category | Activities That the United States Seeks to Deter | Specific Examples |
|---|---|---|
| Aggression, war, conflict globally (conventional)[a] | • Aggression against U.S. interests and formal treaty allies<br>• Aggression against the United States and its territories | • Deter coercion from China toward Taiwan<br>• Deter Russian aggression against Ukraine<br>• Deter Iranian missile threat |
| Attacks against the homeland[b] | • Territorial incursion<br>• Attacks on DoD information systems<br>• Efforts to undermine U.S. or allied democratic systems<br>• Ballistic missile threats and attacks | • None specified |
| Acquisition, proliferation, and use of nuclear weapons[c] | • Nuclear first use<br>• Nuclear terrorism | • Deter China's and North Korea's threat of limited nuclear use<br>• Deter Russia's "limited nuclear first use" |
| Cyberattacks[d] | • Actions against U.S. critical infrastructure<br>• Destabilizing behavior in cyberspace including attacks on U.S. allies | • Deter Russia's cyber operations |
| Attacks in and from space[e] | • Attacks on U.S. space infrastructure<br>• Interference with U.S. ability to operate in space | • None specified |

[a] See, for instance, Jim Mattis, *Summary of the 2018 National Defense Strategy: Sharpening the American Military's Competitive Edge*, Washington, D.C.: U.S. Department of Defense, 2018, pp. 3 and 4; DoD and Office of the Director of National Intelligence, *National Security Space Strategy: Unclassified Summary*, Washington, D.C., January 2011, p. 10; DoD, *Nuclear Posture Review*, Washington, D.C., February 2018c, p. 35; DoD, *Indo-Pacific Strategy Report: Preparedness, Partnerships, and Promoting a Networked Region*, Washington, D.C., June 1, 2019b, introduction by Acting Secretary of Defense Patrick M. Shanahan and p. 44; DoD, *Summary of the 2018 Department of Defense Artificial Intelligence Strategy: Harnessing AI to Advance Our Security and Prosperity*, Washington, D.C., 2018b, p. 12; U.S. Senate Committee on Armed Services, "Statement of Admiral Philip S. Davidson, U.S. Navy Commander, U.S. Indo-Pacific Command, Before the Senate Armed Services Committee on U.S. Indo-Pacific Command Posture," Washington, D.C., February 12, 2019, p. 13 (hereafter, we call this "U.S. Indo-Pacific Command Posture Statement"); Office of the Under Secretary of Defense (Comptroller), *European Deterrence Initiative: Department of Defense Budget, Fiscal Year (FY) 2020*, Washington, D.C., March 2019; Joint Publication 3-0, *Joint Operations*, Washington, D.C.: Joint Chiefs of Staff, January 17, 2017, incorporating change 1, October 22, 2018, p. VII-6.
[b] U.S. Senate Committee on Armed Services, "Statement of General Terrence J. O'Shaughnessy, United States Air Force Commander, United States Northern Command and North American Aerospace Defense Command Before the Senate Armed Services Committee," Washington, D.C., February 13, 2021, pp. 8, I-9. In the NSS, the deterrence concept is extended to illegal immigration in the United States (The White House, *National Security Strategy of the United States of America*, Washington, D.C., December 2017, p. 10).
[c] DoD, 2018c, p. VII.
[d] The White House, 2017, pp. 31–32. See also DoD, *Summary: Department of Defense Cyber Strategy 2018*, Washington, D.C., 2018a, pp. 2, 4.
[e] The White House, 2017, p. 31; DoD, *United States Space Force*, Washington, D.C., February 2019a, p. 4; DoD and Office of the Director of National Intelligence, 2011, p. 10.

This review highlighted several factors relevant to this analysis. First, deterring major aggression against the United States and its allies and partners emerges as the most consistent requirement across many official strategic documents. Nuclear deterrence remains a critical priority, although it is not challenged to the same degree. The most important lesson of U.S. official statements about deterrence, however, is possibly about the growing importance of security issues short of major conflict—activities in the gray zone.[4] In particular, the United States seeks to deter informational, political, and other forms of virtual aggression that constitute a dominant proportion of competitive interactions today. Our analysis therefore had to consider ways in which technology could affect deterrence across a spectrum of major war and competitive actions short of that threshold.

## Criteria for Successful Deterrence: Effectiveness

To assess the criteria that help determine the success of deterrent policies—the efficacy of deterrence, as distinct from the stability of deterrent relationships—we relied primarily on existing RAND work on deterrence, notably the 2018 study *What Deters and Why*.[5] This research included a broad-based survey of the literature on deterrence and several in-depth case studies to generate precisely the sort of criteria for successful deterrence required for our study.[6] Table 3.2 lays out the essential variables derived from that study that we used here to measure deterrence effectiveness.

---

[4] For an analysis of deterrence in the gray zone, see Lyle J. Morris, Michael J. Mazarr, Jeffrey W. Hornung, Stephanie Pezard, Anika Binnendijk, and Marta Kepe, *Gaining Competitive Advantage in the Gray Zone: Response Options for Coercive Aggression Below the Threshold of Major War*, Santa Monica, Calif.: RAND Corporation, RR-2942-OSD, 2019.

[5] Michael J. Mazarr, Arthur Chan, Alyssa Demus, Bryan Frederick, Alireza Nader, Stephanie Pezard, Julia A. Thompson, and Elina Treyger, *What Deters and Why: Exploring Requirements for Effective Deterrence of Interstate Aggression*, Santa Monica, Calif.: RAND Corporation, RR-2451-A, 2018.

[6] Examples of general literature on technology and deterrence include Keir A. Lieber, "The New Era of Counterforce: Technological Change and the Future of Nuclear Deterrence," *International Security*, Vol. 41, No. 4, Spring 2017; Zachary Davis, "Artificial Intelligence on the Battlefield: Implications for Deterrence and Surprise," *Prism*, Vol. 8, No. 2, 2019; and Jaquelyn Schneider, "The Capability/Vulnerability Paradox and Military Revolutions: Implications for Computing, Cyber, and the Onset of War," *Journal of Strategic Studies*, Vol. 42, No. 6, September 2019.

**Table 3.2. Key Variables Governing Effectiveness of Deterrent Threats**

| Category | Variable |
|---|---|
| How motivated is the potential aggressor? | • General level of dissatisfaction with status quo and determination to create a new strategic situation <br> • Degree of fear that the strategic situation is about to turn against them in decisive ways <br> • Level of national interest involved in specific territory of concern <br> • Urgent sense of desperation, need to act |
| Is the defender clear and explicit regarding what it sought to prevent and what actions it would take in response? | • Precision and consistency in the type of aggression that the United States seeks to prevent <br> • Clarity and consistency in the actions that will be taken in the event of aggression <br> • Forceful communication of these messages to outside audiences, especially potential aggressor(s) <br> • Timely response to warning with clarification of interests, threats |
| Did the aggressor view the defender's threats as credible and intimidating? | • Actual and perceived strength of the local military capability to deny the presumed objectives of the aggression <br> • Degree of automaticity of U.S. response, including escalation to larger conflict <br> • Degree of actual and perceived credibility of political commitment to fulfill deterrent threats <br> • Degree of national interests engaged in state to be protected |

Using these foundations, we identified nine criteria for times when technologies could harm the effectiveness of deterrent policies. Given our understanding of the way that deterrence works, in what ways and in what circumstances could an emerging technology—or a set of them— undermine the ability of the United States to effectively conduct its deterrent policies? Table 3.3 outlines those criteria, which form part of the basis for our final assessments, described in Chapter 5.

**Table 3.3. Criteria for Assessing Effect of Technologies on Deterrence Effectiveness**

| Does a given technology . . . |
|---|
| • affect the strategic balance in ways that produce an urgent concern or imperative in a potential attacker that creates a perceived need to go to war? |
| • affect broader strategic balance and geopolitical standing of major powers, future expectations? |
| • provide such decisive and cost-free avenues to aggression that it radically shifts the cost-benefit calculus regarding war, in general or with regard to a specific target/enemy? |
| • offer operational military effects that undermine the credibility of specific deterrent threats? |
| • provide means for regional aggressors to paralyze response from out-of-area allies or sponsors? |
| • provide means to achieve local military goals with great speed? |
| • offer tools and techniques, including nonmilitary means, for engaging in aggression below the threshold of large-scale force but that nonetheless create new facts on the ground-particularly in ways that create misalignment in views, responses between the United States and target of attack? |
| • change the perceived national interests involved in a given potential contingency in ways that reduce the expectation of and/or strategic rationale for a decisive response to aggression? |
| • affect domestic opinion or shift the balance of power among domestic interests in ways that create a domestic political catalyst for aggression independent of deterrent viability? |

One challenge in making this assessment is that meeting one of these criteria alone will probably not be enough for a technology, or a set of several of them, to undermine deterrence. Our assessment had to ultimately identify tipping points where a critical mass of these effects could influence the larger political and strategic judgment behind choices to undertake aggression.

## Criteria for Successful Deterrence: Stability

We adopted a similar approach to developing criteria to judge the effect of technologies on the stability of deterrent relationships. We initially relied on existing RAND work on escalation and stability, created over a range of projects during the past decade.[7] These studies had already examined the literature on strategic stability and escalation dynamics and offered core principles for assessing these characteristics of a strategic relationship. We also surveyed a wider body of literature on escalation[8] and the role of emerging technologies in fomenting instability.[9]

Using these foundations, we identified several criteria for times when technologies could create destabilizing effects in deterrent relationships. We outline those criteria in Table 3.4, which form part of the basis for our final assessments, described in Chapter 5.

---

[7] One leading study was Forrest E. Morgan, Karl P. Mueller, Evan S. Medeiros, Kevin L. Pollpeter, and Roger Cliff, *Dangerous Thresholds: Managing Escalation in the 21st Century*, Santa Monica, Calif.: RAND Corporation, MG-614-AF, 2008. For a more specific discussion of preventive and preemptive escalatory risks, see Karl P. Mueller, Jasen J. Castillo, Forrest E. Morgan, Negeen Pegahi, and Brian Rosen, *Striking First: Preemptive and Preventive Attack in U.S. National Security Policy*, Santa Monica, Calif: RAND Corporation, MG-403-AF, 2006.

[8] David Kinsella and Bruce Russett, "Conflict Emergence and Escalation in Interactive International Dyads," *Journal of Politics*, Vol. 64, No. 4, November 2002; Russell J. Leng, "Escalation: Competing Perspectives and Empirical Evidence," *International Studies Review*, Vol. 6, No. 4, 2004; Barry R. Posen, *Inadvertent Escalation: Conventional War and Nuclear Risks*, Ithaca, N.Y.: Cornell University Press, 1991; and Karen Rasler and William R. Thompson, "Explaining Rivalry Escalation to War: Space, Position, and Contiguity in the Major Power Subsystem," *International Studies Quarterly*, Vol. 44, No. 3, September 2000.

[9] Todd S. Sechser, Neil Narang, and Caitlin Talmadge, "Emerging Technologies and Strategic Stability in Peacetime, Crisis, and War," *Journal of Strategic Studies*, Vol. 42, No. 6, 2019b; Todd S. Sechser, Neil Narang, and Caitlin Talmadge, "Emerging Technologies and Intra-War Escalation Risks: Evidence from the Cold War, Implications for Today," *Journal of Strategic Studies*, Vol. 42, No. 6, 2019a; Erik Gartzke, "Blood and Robots: How Remotely Piloted Vehicles and Related Technologies Affect the Politics of Violence," *Journal of Strategic Studies*, October 3, 2019; Michael C. Horowitz, "When Speed Kills: Lethal Autonomous Weapon Systems, Deterrence, and Stability," *Journal of Strategic Studies*, Vol. 42, No. 6, 2019; Edward Geist and Andrew J. Lohn, *How Might Artificial Intelligence Affect the Risk of Nuclear War?* Santa Monica, Calif.: RAND Corporation, PE-296-RC, 2018; Jürgen Altmann and Frank Sauer, "Autonomous Weapon Systems and Strategic Stability," *Survival*, Vol. 59, No. 5, 2017; and Forrest E. Morgan, *Deterrence and First-Strike Stability in Space: A Preliminary Assessment*, Santa Monica, Calif.: RAND Corporation, MG-916-AF, 2010.

**Table 3.4. Criteria for Assessing Effect of Technologies on Deterrence Stability**

| Does a given technology . . . |
|---|
| • create the potential for successful first-strike options that could degrade the inherent second-strike stability of a nuclear or conventional deterrent relationship? |
| • shift the offense-defense balance toward instability? |
| • create seemingly cost-free routes to aggression—covered in deterrent factors above? |
| • generate the perception that one side in a strategic relationship specifically seeks a form of regime change—the overthrow, destruction, or other radical defeat of its rival in ways that do not accept the legitimacy of its regime or rule? |
| • carry the potential to destroy the sense of mutual equilibrium and agreement to a shared status quo in a strategic relationship through military and geopolitical competition? |
| • create situations in which allies and proxies may intentionally or unintentionally generate destabilizing crises or conflicts? |
| • depend on scarce resources whose control can feasibly become the subject of wars? |

## Challenges to Successful Deterrence: Narratives of Deterrence Failure

We conducted additional research on one aspect of the deterrence challenge. This research was designed to build on existing RAND work by identifying specific routes to deterrence failure. Here, we summarize the analysis that examines how the criteria for effective deterrence (enumerated above) could fail to be met; these criteria informed our assessment of the implications of specific technologies.

Deterrence relies in large part on the apparent strength of deterrent threats (whether of denial or punishment). Deterrence failure can arise when a potential aggressor thinks it can execute a territorial grab or some other form of attack without consequence. But it can also fail when the basic deterrent relationship becomes destabilized—when two or more participants come to believe that they must strike first, for example. Our research therefore sought to assess ways in which technology could undermine either of those pillars of deterrence strength: effectiveness and stability.

A critical fact about deterrence is that there is a distinction between conscious decisions to launch major wars that violate deterrence policies and choices to take much less elaborate actions or provocations—steps that the aggressor believes will *not* lead to war but that end up doing so. The result of the latter is still often categorized as a "failure of deterrence,"[10] but in fact it was not: The aggressor had convinced itself that it could get away with a lesser action without escalation. It can be challenging to identify cases of actual deterrence relationships (as opposed to some other strategic relationship) and then code deterrence failures. Different empirical analyses employ very different data sets of deterrence cases and code these cases in wildly different ways.[11]

---

[10] Paul K. Huth, "Deterrence and International Conflict: Empirical Findings and Theoretical Debates," *Annual Review of Political Science*, Vol. 2, No. 1, June 1999, p. 28.

[11] Richard Ned Lebow and Janice Gross Stein, "Deterrence: The Elusive Dependent Variable," *World Politics*, Vol. 42, No. 3, April 1990.

One clear insight from the literature is that the problem of motivated reasoning and misperception is common across many narratives of deterrence failure. There is not a clear dividing line between these factors; perceptual issues are integral to deterrence dynamics, and the relationship between deliberate and accurate challenges of deterrent threats and aggression through blundering or misperception is more of a spectrum than a dichotomy. Examples of these sorts of perception failures, according to Robert Jervis, include motivated reasoning about the type of conflict that an aggressor can expect, misperceiving the credibility of a defender's commitments and statements, and failing to understand a potential adversary's situation and preferences.[12] As Richard Ned Lebow has argued using multiple historical examples, "Even the most elaborate efforts to demonstrate prowess and resolve may prove insufficient to discourage a challenge when policymakers are attracted to a policy of brinkmanship as a necessary means of preserving vital strategic and domestic political interests."[13] This can complicate any assessment of causal factors responsible for deterrence failure because the problem is sometimes the misperception, rather than accurate interpretation, of actions.

When deterrence has failed through explicit choice, it has typically done so in one of two ways. One way is the decision by one major power (or in some cases multiple major powers) to undertake massive wars of outright conquest, seeking to invade, subjugate, and, in some cases, absorb the national territory of a neighboring or close-by nation-state. This is the failure of central deterrence—states seeking to defend themselves from neighboring revisionist powers—and represents deterrence failure at the hands of a Napoleon or a Hitler,[14] emerging most often as a product of extreme, sometimes irrational urges for expansion or territorial acquisition on the part of the aggressor state. These deterrence failures are sometimes a product of a lack of military preparedness by the defender, which convinces an aggressor that its plans will succeed. But it is also important to stress that highly revisionist powers can become almost undeterrable, so committed to their aggressive ambitions that they will engage in wishful thinking to overcome even significant defensive capabilities.

The second narrative primarily involves failures of *extended* deterrence—situations in which an aggressor (or set of them) has a target state in mind and becomes convinced that the target state's distant ally or sponsor will not come to their rescue.[15] These are often failures of signaling and commitment that allow the aggressor to become convinced that the distant ally or sponsor will stay aloof from the conflict. Two modern cases represent the leading examples of this route

---

[12] Robert Jervis, "Deterrence and Perception," *International Security*, Vol. 7, No. 3, Winter, 1982–1983.

[13] Richard Ned Lebow, "The Deterrence Deadlock: Is There a Way Out?" *Political Psychology*, Vol. 4, No. 2, June 1983, p. 336. See also John Orme, "Deterrence Failures: A Second Look," *International Security*, Vol. 11, No. 4, Spring 1987, p. 97.

[14] Alan Alexandroff and Richard Rosecrance, "Deterrence in 1939," *World Politics*, Vol. 29, No. 4, April 1977; and John J. Mearsheimer, "The German Decision to Attack in the West, 1939–1940," in *Conventional Deterrence*, Ithaca, N.Y.: Cornell University Press, 1983.

[15] On the role of uncertainty and credibility in deterrence see Paul K. Huth, *Extended Deterrence and the Prevention of War*, New Haven, Conn.: Yale University Press, 1988, pp. 1–14.

to deterrence failure: the North Korean invasion of South Korea in June 1950, sponsored by the Soviet Union, and Iraq's 1990 invasion of Kuwait. Farther back in history, World War I also displayed elements of classic failures of credible signaling. In these cases, "leaders plunged their countries into international crises on the mistaken assumption that they could tread on another's commitment without provoking an effective response"—not because of irrational levels of wishful thinking, but because, in each case, "there were in fact weaknesses in the commitment, credibility, or capability of the defender sufficient to tempt an aggressive, perhaps risk-prone, but not necessarily irrational opponent."[16]

Whatever the form of deterrence threat being issued, the issue of credibility is hardly a simple or linear relationship. As Richard Ned Lebow found when looking at cases of provocation and brinkmanship in which an aggressor challenged deterrent threats, "most brinkmanship challenges were initiated without any good evidence that the adversary in question lacked the resolve to defend his commitment."[17] The key factor was not the objective reality but the perceived reality: "What counts is the perception by the initiator that a vulnerable commitment exists—a judgment, we discovered, that was erroneous more often than not."[18] Other recent studies call these findings into question and reemphasize the importance of credibility to calculations of deterrence. Some historical cases of aggressor perceptions—such as statements by Argentine leaders leading up to the Falklands War—make clear that other events convinced them that Great Britain would not respond to an attack.[19]

Finally, an overwhelming lesson of the cases reviewed for this analysis is that the judgment to go to war is a *political* one, involving aspects well beyond the strict cost-benefit calculus of the local military balance and certainly encompassing more issues than the relative balance of technologies. "History indicates that wars rarely start because one side believes it has a military advantage," Lebow writes—or at least not for that reason alone. "Rather, they occur when leaders become convinced that force is necessary to achieve important goals."[20] Indeed, RAND historians and political scientists informally surveyed as part of this analysis could not bring to mind a *single* case of central or extended deterrence failure attributable to the effects of one technology, or a group of them, alone. All these factors conditioned our assessment of the causal relationship between technologies and deterrence.

---

[16] Orme, 1987, p. 121.

[17] Lebow, 1983.

[18] Lebow, 1983, p. 335. Other studies calling into question the simple, linear relationship of credibility to deterrence include Ted Hopf, *Peripheral Visions: Deterrence Theory and American Foreign Policy in the Third World, 1965–1990*, Ann Arbor, Mich.: University of Michigan Press, 1994; Jonathan Mercer, *Reputation and International Politics*, Ithaca, N.Y.: Cornell University Press, 1996; and Daryl G. Press, *Calculating Credibility: How Leaders Assess Military Threats*, Ithaca, N.Y.: Cornell University Press, 2005.

[19] See, for example, Alex Weisiger and Keren Yarhi-Milo, "Revisiting Reputation: How Past Actions Matter in International Politics," *International Organization*, Vol. 69, No. 2, 2015.

[20] Richard Ned Lebow, "Windows of Opportunity: Do States Jump Through Them?" *International Security*, Vol. 9, No. 1, Summer 1984, p. 149.

## Competitor Views of Deterrence

Finally, we examined Russian and Chinese conceptions of deterrence and what role they might play in shaping the effect of emerging technologies on deterrence. Russian and Chinese national security thinking is characterized by persistent ideas about deterrence and stability that differ in some respects from prevailing U.S. ideas. Moreover, each of those countries' security establishments views particular technologies as especially threatening to deterrent relationships. In Chapter 5, we discuss what formed the basis for one of the major perspectives that the project brought to the core research question.

Together, these multiple lines of analysis produced two leading frameworks to help assess the effects of emerging technologies—one on the effectiveness of deterrent policies and the other on the stability of deterrent relationships. In parallel with this analysis, project members conducted deep-dive analyses of each of the eight focus technologies to provide the basis for assessing their potential effect on these criteria. The next chapter summarizes the lessons of these deep dives.

# 4. Overview of Key Technologies

For each of the eight technologies we selected for further examination, we conducted an in-depth analysis that produced descriptions of the technology (including definitions and scope for certain groupings of technologies), the current status of the technology, projected near-term and long-term national security applications, and the potential implications for deterrence. This chapter summarizes the key points that emerged from these analyses.

In this chapter, we focus on the definitions of these technological areas to provide the necessary background for our subsequent analysis of the effects of these technologies on deterrence. We do not delve as deeply here into the potential implications of these technologies on deterrence and national security because Chapter 5 focuses on that. As a reminder, the eight technologies we selected for deeper analysis are (1) advanced cyber and EW, (2) biotechnology, (3) DSSs and technologies, (4) directed energy, (5) hypersonic systems, (6) information- and perception-manipulation technologies, (7) quantum information and sensing systems, and (8) robotics and (semi)-autonomous systems.

## Definitions of Technologies

This first section provides abridged definitions of each of the eight technology areas, including a basic description of the technology, the subcomponents for each technology area, and a discussion of how we scoped each technology for purposes of our study. These sections also offer a brief summary of the potential military or national security applications of each technology.

### Advanced Cyber and Electronic Warfare to Disrupt Networked Information Systems

*Cyber warfare* does not have a universal definition but is generally defined as war that is conducted in the virtual domain, made up of "actions by a nation state or an international organization to attack and attempt to damage another nation's computers or networks for the purposes of causing damage or disruption."[1] Joint Publication 3-12 similarly defines a *cyberspace attack* as "actions taken in cyberspace that create noticeable denial effects (i.e., degradation, disruption, or destruction) in cyberspace or manipulation that leads to denial that appears in a physical domain, and is considered a form of fires."[2] Cyberattacks can fall into two main categories: strategic cyberwar, which consists of attacks launched "for the purpose of affecting [a] target state's behavior," or operational cyberwar, which comprises "wartime

---

[1] Michael Robinson, Kevin Jones, and Helge Janicke, "Cyber Warfare: Issues and Challenges," *Computers and Security*, Vol. 49, 2015.

[2] Joint Publication 3-12, *Cyberspace Operations*, Washington, D.C.: Joint Chiefs of Staff, June 8, 2018, p. GL-4.

cyberattacks against military targets and military-related civilian targets."[3] The Tallinn Manual 2.0[4] defines *electronic warfare* as "the use of electromagnetic [EM] or directed energy to exploit the electromagnetic spectrum."[5] In Joint Doctrine 3-85, EW is called "electromagnetic warfare" and is defined as "military action involving the use of electromagnetic and directed energy to control the electromagnetic spectrum or to attack the enemy."[6] EW is classified into EW-support measures for detection, interception, and neutralization of threats, electronic countermeasures, and electronic counter-countermeasures.[7]

Both cyberspace operations and EW have had varied historical and operational applications in the U.S. military, with EW being an integral part of international military operations since World War I. Although historical divides between cyber warfare and EW have kept the two environments at least theoretically separate, both are becoming analogous as technical capabilities related to the use of the EM spectrum converge. In other words, cyberspace and EM warfare domains, although traditionally very different in implementation and manifestation, could be considered part of a continuum of conflict or operations.

As the cyber landscape evolves, this operational synergy or the convergence of cyber warfare and EW would affect all areas of future military and civilian operations, including such applications as space operations and systems and position, navigation, and timing systems; wireless networks, 5G, IoT devices, and smart cities; critical infrastructure (such as power grids and reservoirs, banking and financial systems, and information transmission systems); military communication systems; and even UAVs and marine vehicles.[8] Future developments in dual-use technologies, such as IoT, AI, and 5G, could create opportunities for adversaries to disrupt sensor-based critical infrastructure or create an environment for "ubiquitous ISR [intelligence, surveillance, and reconnaissance]."[9]

Cyber warfare and EW overlap with and could be directly affected by emerging technologies, such as AI, quantum computing, and directed energy; for purposes of this report, we discuss these related technologies only to the extent that they amplify the vulnerability or the ability to

---

[3] Martin C. Libicki, *Cyberdeterrence and Cyberwar*, Santa Monica, Calif.: RAND Corporation, MG-877-AF, 2009, pp. 117, 139.

[4] Michael N. Schmitt, ed., *Tallinn Manual 2.0 on the International Law Applicable to Cyber Operations*, 2nd ed., Cambridge, UK: Cambridge University Press, 2017.

[5] Schmitt, 2017, p. 565.

[6] Joint Publication 3-85, *Joint Electromagnetic Spectrum Operations*, Washington, D.C.: Joint Chiefs of Staff, May 22, 2020, p. GL-9.

[7] D. Curtis Schleher, *Introduction to Electronic Warfare*, Dedham, Mass.: Artech House, 1986, p. 6.

[8] We refer here to cyber warfare and EW to attack networked information systems, so the potential meaning of that term is very broad. Any attack on an information system—military or civilian, narrow and specialized or broad and society-wide—falls under this category.

[9] U.S. Government Accountability Office, *Report to Congressional Committees: National Security, Long-Range Emerging Threats Facing the United States as Identified by Federal Agencies*, Washington, D.C., GAP-19-204SP, December 2018.

disrupt other technologies using cyber warfare and EW. Box 4.1 summarizes the use of EW and cyber warfare capabilities in a military context.

**Box 4.1**

In potential military applications, cyber warfare and EW capabilities arguably possess the most-direct and well-demonstrated security effects of all eight technology areas. Information security and EW are among the U.S. military's top priorities and among the most significant efforts of U.S. competitors. The rise of 5G communication networks and an increasingly interlinked IoT is creating new vulnerabilities that U.S. adversaries could exploit. These technologies have potential applications both directly in military operations (e.g., attacking U.S. information and decisionmaking systems) and more broadly (e.g., seeking to cause intense disruption in civilian societies during crises or war to disable an opponent).

## Biotechnology

Karl Ereky first coined the term *biotechnology* in 1919, describing it as "all lines of work by which products are produced from raw materials with the aid of living things."[10] Although the definition has evolved slightly over the years, it remains a broad term for the exploitation of biological processes, organisms, cells, and cellular components to develop new technologies or products. Biotechnology draws from many different disciplines, including molecular biology, bionics, bioengineering, nanotechnology, computer and data sciences, genetics, biochemistry, and others.

One widely used classification system for biotechnology breaks it down into subdisciplines based on common applications. These include the following:

- medicine and human health, such as the development of new medicines, therapies, and treatments for diseases
- industrial processes, such as the creation of renewable energy sources
- agriculture, including development of genetically modified plants to increase crop yields or improve insect resistance
- information-processing capabilities at the intersection of bioinformatics, computer science, and chip technology
- processes in marine and aquatic environments
- nutritional biotechnology, such as the fermentation of alcohol and cheese
- law, compliance, and ethical issues within the field
- applications to weapons and warfare.

Although biotechnology is being used in many ways to combat debilitating diseases, reduce our environmental impact, and generally improve our lives and the planet, this same technology has incredible potential for misuse, particularly because it is one of the fastest-growing commercial sectors. Thus, there is an ever-increasing likelihood that knowledge, skills, and/or

---

[10] M. G. Fári and U. P. Kralovánsky, "The Founding Father of Biotechnology: Károly (Karl) Ereky," *International Journal of Horticultural Science*, Vol. 12, No. 1, 2006.

equipment in this field could be adapted for use as biological weapons. Box 4.2 summarizes the potential military applications of biotechnology.

**Box 4.2**

Biotechnology is advancing so quickly that accurate forecasts are very difficult to make. Nonetheless, biotechnology will likely enhance warfighting materiel and systems, optimizing warfighter health and performance, military medicine, and chemical and biological defense technologies. For example, biosensors could protect ground troops from both seen and unseen threats on the battlefield. In the more distant future, a network of biosensors could augment other sensors and intelligence sources to give commanders a more complete picture of the battlefield. Research in biomaterials, biologically inspired materials, and hybrid materials have the potential to revolutionize the wound-healing process along with the design and function of future systems. Other advances in this field could contribute to miniaturization of devices and the development and optimization of biological energy sources of critical value to military design and operations.

## Decision Support Systems and Technologies

Decision support technologies incorporate research and development (R&D) in AI, systems engineering, and information technology to augment or replace some aspects of human decisionmaking. The use of computers to support decisionmaking was an idea presented as early as 1963, and Scott Morton introduced the term *decision support system* in 1971. A typical DSS uses a rules-based approach to sorting, selecting, and transforming data to classify and make recommendations to the decisionmaker. This approach is well suited for structured problems with little uncertainty. Intelligent DSSs (IDSSs) are an emerging type of DSS that uses AI to tackle these unstructured or semistructured problems. These IDSSs are applicable for decisions with a high degree of uncertainty and seek to represent preferences and beliefs of the decisionmaker, using inference to mimic intuition.

DSSs, enabled by AI, can aid, collaborate with, or replace a human decisionmaker. The way that DSSs affect decisionmaking depends on the type of information processing required for the decision and the role of the human in the decisionmaking process. As uncertainty increases, the information processing moves from structured to unstructured, making humans better suited to such tasks. As automation increases, the human role in the decisionmaking process decreases.

To represent the spectrum of automation and information processing in DSS technologies, our analysis focuses on three primary types of DSS across this range:

1. *Automation of rules-based decisions in low-uncertainty contexts.* This type of DSS speeds up decisionmaking by providing support in the form of data visualization or presenting data analysis results necessary for the human to make the decision.
2. *Collaborative decisionmaking via technologies that assess alternative actions and recommend options to human decisionmakers in planning and tactical contexts.* These technologies provide advice to the human decisionmaker, narrowing down the spectrum of decision options, with the decision itself executed by the human.
3. *Artificial cognitive systems that replace aspects of human decisionmaking for unstructured, strategic decisions in highly uncertain environments.* Human

decisionmaking is being supplanted by artificial cognition that duplicates the ability of a human to use intuition and inference to make decisions under uncertainty.

The larger field of organizational decisionmaking systems includes a wide array of technologies, applications, organizations, and procedures. These three focus areas speak to a narrower range of technologies: AI-driven automated systems that undertake part or all of a decision process in support of or as a replacement for human decisionmakers. Although still broad, that focus area defines a relatively specific set of potential technologies. Box 4.3 summarizes the potential military applications of decision support technologies.

**Box 4.3**

Applications of decision support technology to national security include providing data processing and analytic support, performing complex analysis of alternative courses of action, optimizing planning, and developing strategy. AI is already at work in some task automation and decision support, including assisting intelligence organizations in sifting through data. Decision support that allows real-time analysis and exploration of alternative courses of action could also dramatically change military operations. Although this capability is still in the conceptual phase, its development will become more possible as the complexity of AI systems matures.

Directed-Energy Weapons

DEWs inflict damage on their target by applying energy in the form of EM waves or high-speed particles.[11] Because they transmit energy at or close to the speed of light, and because the lasers amass and concentrate energy from the numerous photons or particles in the beam, DEWs have several important advantages over conventional kinetic weapons: Their effects are felt nearly instantaneously, they require virtually no ammunition, and the cost per shot is generally negligible. They share, however, the important disadvantage of being strongly affected by the atmosphere. Visible and near-visible laser light is affected by air turbulence and clouds (i.e., lasers cannot operate in all weather conditions with equal effectiveness), and particle beams are attenuated by air molecules (so they have somewhat limited range in air).

EM weapons are characterized by their wavelength: Most EM DEWs apply energy either in visible (or near-visible) or microwave regions of the EM spectrum.[12] Free electron lasers, which perturb a beam of high-speed electrons, causing them to emit beams of laser light, are a promising technology given their potential for high power and tunability but are inefficient and require a significant footprint to accelerate electrons.

Particle beam weapons, which fire beams of highly energetic particles at targets, are best characterized by the type of particle they fire; the most important distinction is between

---

[11] Strictly speaking, acoustic weapons could also be considered a type of DEW, but we will restrict our attention here to electromagnetic and particle beam weapons that apply energy at or close to the speed of light.

[12] X-ray lasers have also been proposed. In this case, a nuclear bomb could be used to energize inner-shell electrons in a rod of iron, creating an X-ray laser that would be aimed at strategic ICBMs in their midcourse phase. This proposal, however, would violate the Outer Space Treaty of 1967, which prohibits placing weapons of mass destruction in space, and has not seriously been entertained by any major power.

electrically charged particles (electrons, protons, or ions) and neutral particles (neutrons and light atoms, such as hydrogen). Beams of energetic particles cannot propagate more than a few kilometers in sea-level air, so particle weapons are usually considered in high altitude or space settings. Charged particles are difficult to focus in space because of mutual Coulomb repulsion and interactions with the Earth's magnetic field. However, at high altitudes, charged particles can carve out a plasma channel in the air that stabilizes the beam, potentially extending their range to many hundreds of kilometers. Because neutral particle beams do not experience Coulomb repulsion or deflection by magnetic fields, they are quite viable in a space setting and were studied extensively during the Reagan-era Strategic Defense Initiative.

Although short-wavelength lasers (visible and near-visible) share several characteristics common to all DEWs, it is important to note that they differ from microwave and particle weapons in how they deposit energy onto their targets. Both charged and neutral particle beams inflict damage throughout the volume of their targets because they deposit energy via scattering interactions along their entire track. Microwave weapons, with their long wavelengths, are also capable of penetrating inside a target (although they can be shielded to some extent). Laser weapons, however, interact only with the exterior of their targets, depositing energy directly onto the outermost surface. Countermeasures against them are therefore usually considered to be somewhat simpler to implement for laser weapons compared with particle weapons. Box 4.4 summarizes the potential military applications of DEWs.

**Box 4.4**

Missile defense, anti-satellite, and counter-UAS continue to be the largest drivers of investment for near-term DEWs. However, as solid-state lasers become more powerful, their targets are also becoming larger. Short-range air defense, including counter-UAS operations, is a major focus of such R&D. There are major potential implications for space systems: Lasers are likely to become more commonplace for attacking vulnerable targets, such as satellites, over the next few years; both China and Russia are in the late stages of developing ground-based lasers that can be aimed at satellites in low-Earth orbit. The idea of using particle beam weapons for strategic missile defense continues to attract sporadic investment given its potential, and there is particular interest in using lasers to defend aircraft against missile attacks.

## Hypersonic Systems

Hypersonic weapon systems travel at high supersonic speeds, Mach 5 and above—in theory, to speeds as high as Mach 25—or about 5,000 to 25,000 km/hour. There are currently two types: hypersonic glide vehicles (HGVs) and hypersonic cruise missiles (HCMs). Hypersonic systems have three main elements: launch system (rocket) (scramjet/ramjet), delivery system (HGV, HCM), and a payload (conventional or nuclear). HGVs and HCMs could be launched from ground, sea, or air-based platforms, equivalent to the existing traditional U.S. nuclear triad. HGVs are launched on a rocket and shortly after released in the upper atmosphere. Next, they enter an unpowered glide phase at an altitude ranging between about 40 km and 100 km and glide until they reach their intended target. HCMs, which are a faster version of traditional cruise missiles, are first launched on a rocket and then transition to an air-breathing scramjet engine that powers the rest of hypersonic flight, flying at an altitude of about 20 km to 30 km. Alternatively, HCMs can launch on a dual-mode ramjet that would work as a ramjet until it reaches the appropriate altitude and speed and then transition to a scramjet. A *scramjet* is an air-breathing engine that pulls in oxygen from the atmosphere to combine with a liquid fuel (typically hydrogen) to create the combustion required for hypersonic speeds. A *ramjet* carries the liquid oxygen on board.

Hypersonic weapon systems capable of carrying both nuclear and conventional payloads are currently under development and expected to be operational in the next decade. The United States, China, and Russia are the current global leaders for hypersonic technology and have been investing significant resources to field expendable hypersonic weapons as quickly as possible. These weapons can match the high speeds of traditional ballistic missile systems but differ in that they fly at unusually lower altitudes and are highly maneuverable. The combination of maneuverability and altitude poses a significant challenge to existing ballistic missile defense systems designed to defend against a ballistic missile trajectory. Indeed, hypersonic systems pose two major issues for the defending or target country: (1) They are difficult to intercept because interceptors for HGVs must be hypersonic systems themselves;[13] and (2) they compress and complicate decisionmaking timelines because they can reach targets incredibly quickly and can alter their target right up until the final phase of flight (provided the new target is in range). Box 4.5 summarizes the potential military applications of hypersonic weapons.

**Box 4.5**

The 2018 National Defense Strategy identified hypersonic weapon systems as a key technology to enable the United States to fight and win future wars. Current development falls under several U.S. Navy, U.S. Army, U.S. Air Force, and Defense Advanced Research Projects Agency (DARPA) programs that are intended to provide the ability to quickly strike time-critical targets with a conventional payload. Russia is developing two main hypersonic weapon programs, which it views as critical to penetrating U.S. missile defense systems and maintaining nuclear stability. China also views these programs as essential to preventing a U.S. decapitating attack and as a key way to hold U.S. assets at risk.

## Information- and Perception-Manipulation Technologies

Information- and perception-manipulation technologies cover a wide range of tools designed to distort the perception or beliefs of one individual or set of individuals for the purpose of achieving the perpetrator's desired effect. These technologies are generally enabled by AI and aspects of cyber and rely on processing large amounts of data. In the context of international security, this set of technologies enables adversaries to conduct advanced influence operations. For purposes of this report, we examine four mechanisms through which information can be modified, with the goal of influencing or misleading targeted individuals or groups: (1)

---

[13] Loren Thompson, "To Defeat Hypersonic Weapons, Pentagon Aims to Build Vast Space Sensor Layer," *Forbes*, February 4, 2020.

deepfakes, (2) microtargeting, (3) machine learning–driven programs, and (4) spoofing algorithms.

*Deepfakes* are "realistic photo, audio, video, and other forgeries generated with artificial intelligence (AI) technologies."[14] The word *deepfake* itself is recent, dating back to late 2017. Although forgeries have always existed, AI makes them much more sophisticated and harder to differentiate from a genuine photo or video. Making deepfakes is also relatively cheap and easy, broadening the scope of individuals and organizations that can engage in this activity.

*Microtargeting* requires access to large amounts of detailed information on individuals to identify highly specific audiences that can be targeted by a message tailored to match their profile and increase the relevance of the message being communicated. An important characteristic of such "micro-audiences" thus is not so much size as homogeneity—all members of the audience share one or more characteristics that the sender seeks to exploit.[15] Generally used as an advertising tactic, microtargeting can also be used to make phishing attacks more effective by targeting only the most "valuable" (from the attacker's perspective) individuals in a given company or organization (known as "whaling attacks").[16]

*Machine learning* refers to a process that

> involves statistical algorithms that replicate human cognitive tasks by deriving their own procedures through analysis of large training data sets. During the training process, the computer system creates its own statistical model to accomplish the specified task in situations it has not previously encountered.[17]

More-advanced forms of machine learning are referred to as *deep learning*, meaning the algorithm is able to analyze more-complex forms of data and detect more nuance (for instance, identifying images of a car versus a bus or understanding the sentiment behind a given passage of text). Machine learning is considered to be a subfield of AI because it is the process that enables computers to learn how to complete tasks on their own rather than simply executing commands written by humans. One notable application of machine learning has been the development of bots, which are computer programs designed to emulate human behavior, particularly in online interactions. Other applications include speech recognition, image recognition, robotics, and reasoning.

*Spoofing* refers to a form of interference that seeks to obscure or falsify the true source of information (often through impersonation) or replace a stream of information with false or

---

[14] Kelley M. Sayler and Laurie A. Harris, *Deep Fakes and National Security*, Washington, D.C.: Congressional Research Service, October 14, 2019, updated June 8, 2021.

[15] Tom Dobber, Ronald Ó. Fathaig, and Frederik J. Zuiderveen Borgesius, "The Regulation of Online Political Micro-Targeting in Europe," *Internet Policy Review*, Vol. 8, No. 4, December 2019, pp. 2–3.

[16] See, for example, United Kingdom Government, National Cyber Security Centre, "Whaling: How it Works, and What Your Organisation Can Do About It," webpage, October 6, 2016.

[17] Kelley M. Sayler, *Artificial Intelligence and National Security*, Washington, D.C.: Congressional Research Service, November 21, 2019, p. 2, and Keith D. Foote, "A Brief History of Machine Learning," Dataversity webpage, March 26, 2019.

malicious content. Common types of spoofing include caller ID spoofing, email spoofing, media access control (MAC) or Internet Protocol (IP) address spoofing, and Global Positioning System (GPS) spoofing. Caller ID spoofing is the simplest form of spoofing and occurs when "a caller deliberately falsifies the information transmitted to your caller ID display to disguise their identity."[18] Email spoofing is similar in nature and entails manipulating an email to make it look like it came from a different, trusted source rather than the true sender. GPS spoofing is a more sophisticated form of spoofing that consists of "an intentional intervention that aims to force a GPS receiver to acquire and track invalid navigation data."[19] This type of spoofing works by generating false GPS signals to deceive satellite-based navigation systems—collectively referred to as Global Navigation Satellite Systems—into believing they are located somewhere other than their actual position. Box 4.6 summarizes the potential military applications of manipulation technologies.

**Box 4.6**

Manipulation techniques like the ones just discussed could be used to undermine national will in crisis or war by portraying political or military leaders engaging in embarrassing, illegal, or otherwise reprehensible behavior. They could be used as part of traditional deception and concealment operations or connected to much broader and longer-term efforts to undermine the societal coherence of an adversary.

## Quantum Science and Technology

Quantum science combines elements of mathematics, computer science, engineering, and physical science to study the smallest particles of matter and energy: photons and electrons. Quantum science offers the only truly new model for computing that is theoretically capable of solving problems that classical computers cannot realistically solve. The ability to model complex interactions at the subatomic level could enable transformative, rather than incremental, innovation. Because quantum mechanics can be used to describe everything in the natural world, the potential technologies and capabilities that could be realized from harnessing these principles are theoretically limitless. Consequently, quantum mechanics has been championed as the solution to, among other things, cracking many existing data-encryption codes, creating uncrackable codes, defeating stealth technology, enabling AI and machine learning, making the oceans transparent, creating new materials, and discovering and curing diseases.[20] Yet some applications of quantum principles are better developed than others, and most remain in the

---

[18] Federal Communications Commission, "Caller ID Spoofing," webpage, January 6, 2020.

[19] Elahe Shafiee, Mohammad Reza Mosavi, and Maryam Moazedi, "Detection of Spoofing Attack using Machine Learning based on Multi-Layer Neural Network in Single-Frequency GPS Receivers," *Journal of Navigation*, Vol. 71, No. 1, 2018.

[20] Alexandre Ménard, Ivan Ostojic, Mark Patel, and Daniel Volz, "A Game Plan for Quantum Computing," *McKinsey Quarterly*, February 6, 2020; and, Jason Palmer, "Quantum Technology Is Beginning to Come into Its Own," *The Economist (Technology Quarterly: Here, There and Everywhere)*, 2017.

experimental stage. In general, quantum technologies can be grouped into three categories: sensing and metrology (measurement), communications, and computing.[21]

Although the fields of quantum theory and mechanics have existed since the turn of the 20th century, the actual creation of quantum devices has followed much more slowly. A pacing factor has been the development and refinement of enabling technologies, including highly tuned lasers, semiconductors, and techniques for controlling the environments within which quantum objects can usefully operate.[22] With the maturation of these enabling technologies, quantum technologies in the three main categories listed earlier have been demonstrated in laboratories, deployed at small scales, or commercially deployed. Advancing these classes of technologies to the point of useful defense or commercial applications, however, will still require overcoming numerous technical challenges to achieve the required improvements in reliability, performance, and cost.[23] As they continue to advance, quantum technologies have the potential to be transformative for warfighting, information security, AI, materials science, medicine, geology, and space exploration.[24] Box 4.7 summarizes the potential military applications of quantum science and technology.

**Box 4.7**

The national security effects of quantum science and technology remain more theoretical than in many of the other technology areas. However, if and when they reach fruition, they are expected to have dramatic implications for such military issues as communications and cryptography. Potential applications of quantum sensing—in particular, quantum inertial navigation systems and gravimeters—will likely not reach the point of being sufficiently small, light, low-power, or cost-effective in the near term. Quantum communications and computing remain highly experimental.

## Robotics and Semiautonomous Systems

The category of robotics and semiautonomous systems encompasses several more-specific types of technology. They range from fully autonomous robotic systems to semiautonomous systems that augment human direction with a degree of self-direction. *Robotic* systems are generally viewed as fully autonomous military systems—whether surveillance, communication, strike, or for other purposes—whereas the majority of current platforms in use represent some

---

[21] U.S. Air Force Scientific Advisory Board, "Utility of Quantum Systems for the Air Force," study abstract, 2015.

[22] Palmer, 2017.

[23] National Academies of Sciences, Engineering, and Medicine, *Quantum Computing: Progress and Prospects*, Washington, D.C.: National Academies Press, 2019; Keith W. Crane, Lance G. Joneckis, Hannah Acheson-Field, Iain D. Boyd, Benjamin A. Corbin, Xueying Han, and Robert N. Rozansky, *Assessment of the Future Economic Impact of Quantum Information Science*, Washington, D.C.: IDA Science and Technology Institute, August 2017; and U.S. Air Force Scientific Advisory Board, 2015.

[24] Scott Buchholz, Joe Mariani, Adam Routh, Akash Keyal, and Pankaj Kamleshkumar Kishnani, "The Realist's Guide to Quantum Technology and National Security," *Deloitte Insights*, February 6, 2020.

version of partial autonomy that is either directly piloted or guided in a more general sense by human operators. For this report, we examined a number of categories of such systems:

- **UAV:** A type of aircraft piloted without a human on board.
- **Autonomous UAV:** A UAV that can "sense, communicate, plan, make decisions and act without human intervention."[25] This type of autonomy is called "human out of the loop."
- **Semiautonomous UAV:** A UAV on which some aspects of perception, reasoning, and action are still performed by a human operator. Depending on the extent of operator control, this is either called "human on the loop" or "human in the loop."
- **Remote-piloted aircraft:** A UAV that is either semiautonomous or has no autonomy capability at all.
- **Lethal autonomous weapon systems (LAWS):** DoD Directive 3000.09 defines LAWS as "weapon system[s] that, once activated, can select and engage targets without further intervention by a human operator."[26] A weaponized, autonomous UAV is a LAWS.
- **UAS:** A UAS comprise a UAV, support element (e.g., transportation and maintenance equipment), human element (aircraft pilot and payload operator), control element (e.g., the ground control station), and data link (e.g., communications satellite).
- **Swarm:** A group of UAVs working collectively to accomplish a mission. Concepts of employment for autonomous UAVs range from large swarms of cheap systems to single, expensive, but more-capable ones.

Such systems could have significant effects on the character of military operations and perhaps warfare. The United States and its leading competitors are making significant investments in UAV/UAS technology and plan to use such systems, both semiautonomous and eventually fully autonomous, for a whole range of military missions. Some proposals suggest that massive numbers of very small and cheap drones could work in swarms to overwhelm defensive and self-protective systems. The result could be a future of conflict dominated by the effects of autonomous systems. A complete transition of this sort probably lies beyond the time frame of this report, but unmanned systems are some of the most mature of any of the technology areas we considered, and even a partial emergence of their potential could have dramatic effects on military operations—and pose new risks to deterrence.

For scoping purposes, we chose to focus our analysis on semiautonomous UAVs. The capabilities of a UAV result from integrating multiple technological components: Closed-loop target tracking, for example, requires a combination of sensors, algorithms, and platforms. We did not consider related systems or capabilities, such as data links essential to UAV operations, loitering munitions, ground or undersea autonomous systems, or systems with limited mobility. Box 4.8 summarizes the potential military applications of robotics and semiautonomous systems.

---

[25] Yasmina Bestaoui Sebbane, *Smart Autonomous Aircraft: Flight Control and Planning for UAV*, 1st ed., Boca Raton, Fla.: CRC Press, 2015.

[26] Department of Defense Directive 3000.09, *Autonomy in Weapon Systems*, Washington, D.C., incorporating change 1, May 8, 2017, p. 13; and Kelley M. Sayler, "Defense Primer: U.S. Policy on Lethal Autonomous Weapon Systems," Washington, D.C.: Congressional Research Service, IF11150, 2019.

**Box 4.8**

After cyber warfare and EW, robotic and semiautonomous systems have the most-immediate and perhaps large-scale potential military implications. These systems are already widely used in counterterrorism; ISR; and other roles, and in the future could become the dominant platforms for sensing of all kinds, precision strikes, delivery of cyber warfare or EW packages, and even air-to-air engagements.

## Current Status, Future Trajectory, and Limitations

Having defined the technologies, the deep-dive analyses then considered the current status of the technologies, their projected near-term and long-term development, and the constraints or challenges for development that might limit their effects on deterrence. We have distilled the essential points into the sections below, providing a brief review of the current status and likely prospects of each technology area.

These deep-dive analyses of the eight technology areas—when combined with the assessment of the character of deterrence—set the stage for the final major analytical phase of our study. This phase helped us to assess the ways that these technologies could affect the effectiveness and stability of deterrence. The next chapter offers our approach and findings on that issue.

### Advanced Cyber and Electronic Warfare

The bulleted list summarizes the current status, future trajectory, and challenges and constraints associated with advanced cyber warfare and EW.

- **Current capabilities:** Advanced cyber warfare and EW have the capacity to disrupt critical networked information systems. Targets include 5G and IoT; space systems and positioning, navigation, and timing; GPS and sensing systems; and blockchain and distributed ledger technologies. Forms of attacks include spoofing, tampering, and GPS jamming.
- **Near-term national security applications:** The next five to ten years may see the advent of a fully functional 5G communication network that would facilitate an exponential use of IoT devices in military applications, offering pervasive sensing, communication, and analytics for improved situational awareness (e.g., internet of military things, internet of battlefield things). Military use of these technologies could create threats to the safety of communications/information, mission operations, personnel, and even supply chains.
- **Long-term national security applications:** Space could see proliferation of constellations, particularly nanosatellite constellations, that could boost resilience and expand coverage areas. With the proliferation of IoT devices, sensors, and networks, the need to manage and secure IoT environments could be fulfilled by the use of blockchain technology.
- **Challenges or constraints for development:** Advances in critical networked information systems also generate new vulnerabilities; the increasingly networked nature of systems (both civilian and military) makes them vulnerable to cyberattacks and EW,

which can slow the pace of new developments, because these vulnerabilities must be addressed before further progress can be made.

## *Biotechnology*

In this section, we summarize the current status, future trajectory, and challenges and constraints associated with biotechnology.

- **Current capabilities:** Common applications are medical (e.g., antimicrobial resistance), agricultural (e.g., genetically modified organisms), and industrial (e.g., biofuels) in nature.
- **Near-term national security applications:** Major near-term applications include biosensors (detect harmful substances), molecular electronics (high-speed signal processing and communication, volumetric data storage), materials (improves wound healing capabilities and form, fit, and function of battlefield equipment), logistics (miniaturization and the development and optimization of biological energy sources), and therapeutics.
- **Long-term national security applications:** Scale, scope, complexity, and tempo of products are likely to increase, and the character and type of the actors involved will be even more diverse. In the more-distant future, biosensors could augment other intelligence sources to give a more complete picture of the battlefield.
- **Challenges or constraints for development:** For biosensor technology to come to bear, both sensitivity and specificity need to increase. Sensors also need to be smaller, more portable, and capable of withstanding harsh environments.

## *Decision Support Systems and Technologies*

In this section, we summarize the current status, future trajectory, and challenges and constraints associated with DSSs and decision support technologies.

- **Current capabilities:** Augmentation technologies can increase the speed and efficiency of decisionmaking and of data reporting, analysis, and interpretation. Examples include DARPA's Target Recognition and Adaptation in Contested Environments (TRACE). program, Elon Musk's Neuralink (a brain-computer interface startup), and IBM's Watson.
- **Near-term national security applications:** Applications of decision support technology to national security include providing data processing and analytic support, performing complex analysis of alternative courses of action, optimizing planning, and developing strategy. Combining an increased ability to sort and analyze data with developments in models of human behavior could improve decisionmakers' ability to make strategic predictions.
- **Long-term national security applications:** Replacing strategic decisionmaking with automated artificial cognition is a distant possibility.
- **Challenges or constraints for development:** A critical limitation to machine learning is that this data-driven approach relies on the quality of the underlying data and is therefore inherently brittle, sensitive to biases in the data and spoofing, and poor at recognizing novel events or operating in an environment with high uncertainty.

*Directed-Energy Weapons*

In this section, we summarize the current status, future trajectory, and challenges and constraints associated with DEWs.

- **Current capabilities:** DEWs are still relatively limited in their military scope. Lasers are not yet powerful enough to contest hardened military targets, and particle weapons have not yet been fielded by any nation outside of limited applications in strategic missile defense. High-power microwave weapons have been used with success against various targets. DEWs have also been employed in nonlethal anti-personnel roles (e.g., for crowd control) with moderate success.
- **Near-term national security applications:** Missile defense, anti-satellite, and counter-UAS continue to be the largest priorities for near-term DEWs. Short-range air defense against rotary- and fixed-wing aircraft (both manned and unmanned) is a goal of several Army programs. Solid-state lasers are becoming more powerful and will likely be used more frequently to attack vulnerable targets, such as satellites. The focus of DEWs today is on tactical rather than strategic uses, but particle beam weapons may be used for strategic missile defense in the near future.
- **Long-term national security applications:** Improvements to solid-state laser technology, especially in power and efficiency, may make lasers effective against armored targets, but this requires years of research. In the more distant future, electron beam weapons may enjoy renewed interest as plasma wakefield acceleration technology matures.
- **Challenges or constraints for development:** In their present form, relatively low-power laser weapons still require a large amount of energy, which is limited even on a warship. As a result, the military applications of these weapons are currently restricted.

*Hypersonic Systems*

In this section, we summarize the current status, future trajectory, and challenges and constraints associated with hypersonic systems.

- **Current capabilities:** Expendable HGVs and HCMs are still in the development phase, with HGVs being farther along than HCMs. Recent tests include the Common-Hypersonic Glide Body (C-HGB) in March 2020. Russia claimed that it had attained the first deployable HGV at the end of 2019; the United States currently expects to have a deployable system in 2022.
- **Near-term national security applications:** The 2018 National Defense Strategy identified hypersonic weapon systems as a key technology to enable the United States to fight and win future wars. Presently, DoD has only funded operational prototypes of hypersonic weapon systems but has not yet decided to acquire them.
- **Long-term national security applications:** Development and deployment of expendable HGVs is expected for several global entities. HCMs will be slower to develop than HGVs because they rely on a scramjet. A potential longer-term development could be the integration of HCMs into systems that are both reusable and potentially manned.

- **Challenges or constraints for development:** There are four key technical barriers for HGVs and HCMs: propulsion, thermal management and materials, flight control, and testing, modeling, and simulation in the hypersonic regime.

## Information- and Perception-Manipulation Technologies

In this section, we summarize the current status, future trajectory, and challenges and constraints associated with information- and perception-manipulation technologies.

- **Current capabilities:** Deepfakes are gaining in sophistication; they are hard to detect without the use of special detection software. Microtargeting makes it possible to direct messages to individuals who match specific criteria. Telephone and email spoofing have long been in widespread use because of their low cost and ease of execution. GPS spoofing is following suit because its enabling software-defined radio technology is becoming less costly and more accessible. As these technologies mature, however, so do detection capabilities.
- **Near-term national security applications:** The use of deepfakes could trigger destabilizing shifts in public opinion or large-scale popular reactions and an increased level of uncertainty in decisionmaking. Microtargeting enables the targeting of specific individuals in the military with false information campaigns or threatening messages. Machine learning allows for enhanced data collection and analysis capabilities, and the development of more-precise targeting systems but can also facilitate information warfare and influence operations. GPS spoofing can enable adversaries to capture U.S. equipment, mislead U.S. assets, or obscure their facilities or movements to make it difficult for the United States to target them.
- **Long-term national security applications:** Deepfakes are expected to become increasingly sophisticated and more widely available to the general public. Ability to microtarget individuals is growing. Spoofing technologies are projected to become more prolific and sophisticated in the long term, although anti-spoofing technologies are attempting to keep pace.
- **Challenges or constraints for development:** Of this set of technologies, machine learning as applied to information- and perception-manipulation faces inherent limitations: Algorithms are dependent on the training data sets and are thus prone to replicating errors and may not know how to handle information not contained in the data set. Algorithms can also be manipulated into learning incorrect information, making them vulnerable to exploitation by adversaries.

## Quantum Information and Sensing Systems

In this section, we summarize the current status, future trajectory, and challenges and constraints associated with quantum information and sensing systems.

- **Current capabilities:** Several quantum technologies (e.g., chip-scale magnetometers, chip-scale atomic clocks, quantum inertial navigation systems) have been demonstrated in laboratories, deployed at small scale, or commercially deployed. Quantum metrology and sensing technologies have demonstrated the highest level of military and commercial utility and readiness in terms of quantum application development. Quantum computing, on the other hand, remains highly experimental, and numerous technical challenges must be addressed before a usable quantum computer can be fielded.

- **Near-term national security applications:** In terms of computing, fielded machines are rudimentary, and existing quantum computers largely establish a proof of principle; software and algorithms are in similarly nascent stages. The most heralded potential quantum applications include cryptography and cryptanalysis, enabling precision timing and navigation in GPS-denied environments, identification of moving masses underwater and underground structures, the end of stealth, and pinpointing electric field sensors and communication receivers. In the next decade, however, few if any of these technology categories are likely to advance to the point where they are capable of being deployed on combat systems or in support of combat operations.

- **Long-term national security applications:** Quantum computing hardware by itself will not be useful without the software and algorithms to query the machine; researchers estimate it will take eight to ten years of investment to build and demonstrate a large-scale, fault-tolerant, gate-based quantum computer from the point at which a system design plan is finalized. Presently, no such plan exists. Quantum inertial navigation systems and gravimeters will likely not reach the point of being sufficiently small, light, low-power, or cost-effective in the near-term.

- **Challenges or constraints for development:** Several quantum technologies currently face constraints in their size (i.e., they are too large to be placed on military aircraft), accuracy, or cost to produce. Others face inherent constraints; for instance, quantum communications cannot make data transfer faster because nothing can move faster than the speed of light. Other technical challenges include vulnerability to hacking, issues with storing massive quantities of information, and access to required space-based assets.

## Robotics and Semiautonomous Systems

In this section, we summarize the current status, future trajectory, and challenges and constraints associated with robotics and semiautonomous systems.

- **Current capabilities:** Current military UAVs are characterized by a wide diversity of advanced platforms and payloads but limited autonomy algorithms. Current types of UAVs with some level of autonomy include the following, with specific examples of systems in parentheses: medium altitude, long endurance (MALE) ISR and air-to-ground strike (MQ-9 Reaper); high altitude, long endurance (HALE) ISR (RQ-4 Global Hawk); and low-altitude ISR and payload delivery (FLIR SkyRaider).

- **Near-term national security applications:** Semiautonomous UAVs are currently employed in counterterrorism and counterinsurgency operations and as enablers of permissive air environments. As of 2019, the USAF maintains 70 MQ-9s airborne on a nearly continuous basis. Strategic capabilities expected to be developed in the next five to ten years include nonlethal support for manned aircraft; collaborative swarm ISR and assisted strike in denied airspace; and long-range strategic bombing and/or ISR in denied airspace.

- **Long-term national security applications:** Projected strategic capabilities expected to be developed in the long term include the ability to more easily distinguish between friend and foe in air-to-air combat (loyal wingman platforms) and faster, larger-scale collaborative ISR and assisted strike with swarms in denied airspace. A range of autonomous unmanned vehicles for the ground, surface and undersea, and space domains are also in development.

- **Challenges or constraints for development:** Currently, autonomy in most platforms is restricted to various forms of waypoint flight. Additionally, presently, several platforms (e.g., MQ9 Reaper and RQ-4 Global Hawk) have low survivability in denied or contested environments. UAVs that rely on AI-enabled DSSs may also face some of the same challenges related to fidelity of targeting. In some contexts, the use of UAVs may face ethical or legal constraints, which could jeopardize or limit R&D budgets for such systems.

# 5. Effects of Technologies on Deterrence

The final step in our research was to take the products of the two major previous phases of analysis—the review of deterrence requirements and the deep-dive appraisals of the technologies—and identify possible causal effects of those technologies. This was the core analytical step of the project: to conduct rigorous analysis to identify ways that our selected technologies could shape the effectiveness and stability of deterrence.[1]

As noted in the introduction, such findings will inevitably involve qualitative judgments rather than the outputs of models or other quantitative analysis. Identifying such causal relationships posed many methodological challenges. There are many variables involved in this causal relationship, and no methodologically sound mechanism for isolating specific effects of new technologies. Existing large-$N$ quantitative studies on deterrence, for example, only test for truly major factors, such as overall military balance and strength of commitment, and even these studies have questionable success isolating variables from one another.

Deterrence, moreover, hinges entirely on perceptions. Even if a given technology could objectively affect a military relationship, unless it also shapes the perceptions of decisionmakers, it will not have a parallel effect on deterrence. Indeed, general military effects of a technology are not the same as effects on deterrence; existing analyses of the effect of technology on war outcomes or military effectiveness, for example, are largely irrelevant unless we can demonstrate that these capabilities shape perceptions.

Nor are the causal effects of technologies likely to be simple or linear. There are counters to any technology that could dilute its effects; assessing the net effect can be difficult given such interactive dynamics. Moreover, any given technology can be used to strengthen as well as undermine deterrence. To further complicate matters, some technologies are too costly, meaning that even if their capabilities could theoretically undermine deterrence, in practice, no state will be able to deploy enough of the technology to have such an effect.

Moreover, the literature on the military operational effects of technology speaks to the limits of individual or even collective sets of technologies to decisively shift military outcomes on their own, as well as the challenge of generating sufficient military innovation in peacetime to produce revolutionary advances.[2] This literature reinforces a general finding of this research—

---

[1] This chapter summarizes the detailed analysis of each of these assessment criteria.

[2] See Stephen Biddle, *Military Power: Explaining Victory and Defeat in Modern Battle*, Princeton, N.J.: Princeton University Press, 2006; Michael Horowitz, "Do Emerging Military Technologies Matter for International Politics?" *Annual Review of Political Science*, Vol. 23, May 2020; Michael C. Horowitz, *The Diffusion of Military Power: Causes and Consequences for International Politics*, Princeton, N.J.: Princeton University Press, 2010; Keir A. Lieber, *War and the Engineers: The Primacy of Politics over Technology*, Ithaca, N.Y.: Cornell University Press, 2005; and Stephen Peter Rosen, *Winning the Next War: Innovation and the Modern Military*, Ithaca, N.Y.: Cornell University Press, 1991.

that emerging technologies *alone* are unlikely to have decisive effects on deterrence. We took this qualification seriously in our analysis, looking initially to see if specific emerging technologies had the potential to break this historical pattern and then considering their effects in a larger context. We chose one lens in our application analysis, which highlighted the role of technologies working in combination rather than alone, specifically influenced by the basic insight of this related literature.

To deal with such complexities and ground our judgments in the most rigorous analytical process possible, we took several steps to bring discipline to those judgments. We first had the technology deep-dive authors propose ways in which their technologies might affect deterrence, independent of the deterrence framework developed in the project (and laid out in Chapter 3). Second, we developed that framework to test the effects of the technologies against specific criteria. Those two steps then flowed into the primary analytical step: We formed separate working groups to assess the effects of technologies on deterrence in that framework through four distinct lenses. These analytic lenses were: (1) the effect on deterrence credibility, (2) the effect on deterrence stability, (3) the implications of the ways that U.S. competitors view deterrence, and (4) the ways that combinations of technologies could affect deterrence.

For each lens, we looked for five specific forms of causal effects. The analytical teams looked first for straightforward and obvious potential causal effects: for example, the potential of a strike system to make certain targets vulnerable. They then looked for a repeated, common pattern of potential effects across multiple technologies or for one technology across multiple forms of deterrence. They also sought to identify potentially dramatic or decisive effects—technologies that could paralyze a defender and undermine deterrence in decisive ways. The teams then sought to identify historical or current examples of the technology already having such effects. Finally, they reviewed the project's analysis of Russian and Chinese views of these technologies for evidence that U.S. rivals seek to exploit these causal effects.

This analytic process enabled us to nominate important ways in which these technologies *may* affect deterrence. We cannot claim that this analysis identified all such causal connections, nor can it be said to prove that the technologies will certainly have these effects. As this chapter and the following one will explain, however, this analysis does identify potentially dangerous effects of these technologies on deterrence effectiveness and stability and points to clear implications for DoD and USAF policies. This chapter summarizes the results of each of the four analyses. Chapter 6 then offers general themes and conclusions that are derived from the work as a whole.

## Lens 1: Effects on Deterrence Credibility

The first analytical team looked at possibly the most straightforward of the questions: How the eight focus technologies might affect the credibility of deterrent threats, primarily by focusing on how technology influences military capability. (This criterion speaks to the first broad focus of the study—the effectiveness of deterrence.) In some instances, technological developments may alter the ability of a deterrer to conduct an effective defense and inflict battlefield damage on an attacker (deterrence by denial). If technology makes an attack more difficult or costly, then deterrence will likely be strengthened. If, however, technology provides a meaningful advantage to an attacker, then deterrence may be weakened or destabilized. In other cases, new technologies might alter a deterrer's ability to punish the challenger for an attack by striking at critical strategic countervalue targets (deterrence by punishment). Technologies that are hard to defend against are likely to strengthen deterrence; conversely, those that offer the prospect of a successful and overwhelming disarming first strike are likely to weaken or destabilize it.

If both sides have similar levels of technological capability and are peer or near-peer rivals, various broad outcomes are theoretically possible. One possible outcome is that the technological capabilities "cancel each other." In other words, any strategic or tactical advantage that one side might gain is erased by the fact that the other side now has the same advantage. For instance, in the conventional domain, stealthier weapons on both sides would make it more likely that the challenger can evade the defender's battlefield defenses and strike at critical operational targets, such as command and control nodes, armored columns, or logistics centers, thus increasing the probability of a battlefield success (weakening deterrence by denial). However, stealthier weapons would also make it more likely that the defender can evade the challenger's own battlefield defenses, thus enabling them to strike at targets that cause the offensive to grind to a halt or to strike at strategic targets that damage the aggressor's economy, political system, or societal control and punish it with devastating strikes against important strategic targets (strengthening deterrence by punishment). These technologies do not necessarily need to be symmetric in nature; because one side may be able to reduce the effectiveness of a rival's technology with a different set of technologies of its own.

In other cases, however, technology might have asymmetrical effects that favor either the defender or the challenger. An example of this might be the deployment of an operationally effective DEW for use in conventional scenarios to attack space-borne systems and other military capabilities. Such systems, at least initially, could provide a critical defensive advantage by significantly reducing the effectiveness of airborne and other long-range strike or ISR systems. Under such conditions, an attacker would need to rely on potentially costly ground attacks conducted with degraded information about the defender's location and disposition. By making offensive operations more costly, DEWs should enhance deterrence at least until successful counters can be developed.

## Factors Mitigating Technological Impacts on Deterrence

Although the different effects have the potential to undermine deterrence, there are also some important mitigating factors to take into consideration. One is the recognition that a single factor will rarely break deterrence on its own. For instance, scholars have shown preventive wars to be relatively rare; few countries choose to start a war either out of fear that their relative position will deteriorate after their adversary acquires a given technology or just to take advantage of a window of opportunity. Of more relevance is how the combination of such effects might act as a tipping point, in the context of larger political and strategic judgment, to undermine deterrence and provoke war.

Cost represents another important limitation to what technology can achieve when it comes to deterrence. A technology might be too expensive to deploy in large numbers, limiting the incentive to wage a war of opportunity as described above. A technology might also carry other types of costs, such as the moral cost of using it for the policymakers who would make that decision.

It is also important to keep in mind that an adversarial relationship is a dynamic process, in which one side responds and adapts to the other. In this context, deterrence outcomes depend on lasting and meaningful differentials in development of the technologies. A technology may not have a deterrent effect if the states in question have the same capabilities and the operational advantages are minimal. Conversely, a technology could increase the probability of escalation if it provides a significant military advantage to a rival who also is dissatisfied with the current status quo.

Two other factors may influence how technologies undermine or reinforce deterrence: the transient nature of any technological advantage gained and the degree to which the military effects of a technology are understood and have been operationally demonstrated. It is difficult—and historically rare—for a state to maintain a lasting technological advantage over a peer or near-peer rival. The action-reaction cycle inherent in many technologies tends to render a technological advantage transient or mitigate its operational effects.[3] The effect of technological transience on deterrence may also depend on the risk acceptance of the relevant decisionmakers: During a crisis, a risk-acceptant leader may be more willing to exploit a perceived temporary technological advantage, think that it is easier to convince that such an advantage is militarily relevant, or is more likely to act out of fear that a window of opportunity is closing.

---

[3] Williamson Murray's *Military Adaptation in War: With Fear of Change* (Cambridge, UK: Cambridge University Press, 2014) describes a dynamic, borne out by history, in which states compete with one another to gain technological advantages in battle, which are usually fleeting and may not always yield the desired results. One reviewer of Murray's work summarized his arguments surrounding this dynamic, stating:

> Advances in the direct and indirect technologies of warfare also tend to increase the speed of military operations and the potential scope of their destruction. Every weapon and every military system introduced eventually spawns a countersystem. Change, then, is a constant in warfare; and as technology advances, the tempo of that change increases. The side that manages to stay at least one step ahead of the enemy usually has the advantage. But not all change introduced by armies works or turns out to be change for the better. Therein lies the rub. (David T. Zabecki, "Book Review: Military Adaptation in War, by Williamson Murray," HistoryNet webpage, undated)

The second factor affecting how a technology influences deterrence stability focuses on the perception of a technology's operational utility.[4] The effects of new technologies, particularly nonkinetic ones, can often be speculative until they are deployed operationally in a combat environment or are integrated into combat formations. This lack of a well-grounded understanding of the operational effects of a new technology means that perceptions of a technology's utility can drive deterrence calculations and may differ across decisionmakers.

This is a significant consideration, and we tried to take account of it in our assessment of the implications of specific technology. Because deterrence is ultimately a perceptual game of sorts, the objective effects of a new technology—although they may be the dominant consideration for that technology's military operational effects—are less important than what each side *believes* their effect is likely to be. If China saw a given new U.S. technology as decisive (but not threatening to its homeland or indicative of a U.S. desire to strike first), that technology would bolster deterrence, even if the Chinese perception was inaccurate. Assessing the effects of technologies in deterrence is primarily a process of understanding their perceptual effects.

## Assessing the Effects on Deterrence Credibility

To assess the possible implications of our eight technology areas for these factors, we considered the five classes of causal effects noted earlier in this chapter: straightforward, undeniable, and potentially significant effects on deterrence; a repeated, common pattern of potential causal links; cases of potentially dramatic or decisive effects; historical or current examples of the technology already having the effect; and specific evidence that U.S. rivals intend to use the technology to have the purported effect on deterrence.

Technologies can affect the credibility of deterrence in many ways. They can have a direct effect on combat outcomes—weapons that can be used on the battlefield to destroy, damage, or render ineffective an opponent's military capability (deterrence by denial) or strategically to inflict harm on an opponent's population, economic capacity, or political system (deterrence by punishment). Even if they do not have such direct combat effects, they can enable or amplify the effects of existing military tools—systems that integrate existing military capabilities into more effective capabilities, improve command and control, enhance ISR, or improve sustainment and logistics. More indirectly, they can shape the strategic environment in which military force is used. Technologies can have such effects as providing new ways of influencing decisionmakers and populations, increasing economic productivity that can be used to support military R&D or production, introducing new economic or societal vulnerabilities, creating new resource dependencies, or shifting the overall balance of power.

---

[4] Adam Grissom argues that a development is only considered to be a "military innovation" if it constitutes "'a change in operational praxis that produces a significant increase in military effectiveness' as measured by battlefield results" (Adam Grissom, "The Future of Military Innovation Studies," *Journal of Strategic Studies*, Vol. 29, No. 5, 2006, p. 97).

Our analysis identified a range of specific potential technological effects on the credibility of deterrence. It highlighted five leading routes to deterrence failure under the influence of emerging technologies, summarized in Table 5.1.

**Table 5.1. Potential Effects on Deterrence Credibility**

| Potential Effects | Technologies That Promise or Threaten Such Effects[a] |
|---|---|
| Create decisive military operational effects from speed of operations (initial and iterative cycles). | • Cyber/information attack<br>• DSSs<br>• Hypersonics<br>• Unmanned systems (AI empowered)<br>• Quantum technologies (longer term) |
| Make the defender "blind"; decisive impact on awareness; command, control, communications, computers, intelligence, surveillance, and reconnaissance (C4ISR). | • Cyber/information attack, systems warfare<br>• Information and perception manipulation (spoofing, deepfakes)<br>• Hypersonic strikes on critical nodes<br>• Swarming unmanned concepts<br>• Quantum technologies (longer term) |
| Acquire more-perfect battlefield awareness (pervasive ISR). | • Unmanned loitering, persistent systems<br>• Open-source networked data (IoT, imagery)<br>• Biotech-based sensors |
| Manipulate opinion, attitudes, and national will. | • Information- and perception-manipulation technologies<br>• Biotechnology |
| Empower aggression below the threshold of war. | • Cyber/information attack<br>• Information- and perception-manipulation technologies<br>• Unmanned systems<br>• Biotechnology (gradual employment) |

[a] This list reflects both the general eight categories of technology examined in this report and, where we saw value in highlighting subcategories, more-specific technologies or capabilities that are component parts of those eight broad technology areas.

Of these, one possible technological effect emerged as the most potentially significant of all: the employment of a combination of technologies (cyber and information warfare, EW, information and perception manipulation, AI/machine learning, decision support, and even UAS to deliver information or EW packages) to shock and effectively paralyze the enemy's command, control, and warfighting system. The concept is central to Chinese warfighting doctrine ("system destruction warfare") and a major focus of Russian operational art as well.[5]

---

[5] This idea, in fact, parallels a notion in modern French military doctrine: As Michael Shurkin ("Modern War for Romantics: Ferdinand Foch and the Principles of War," *War on the Rocks*, July 8, 2020) explains,

*Foudroyance*, derived from the word for thunder (*foudre*), means a sudden crippling shock.

He quotes a French military thinker, Admiral Guy Labouérie, as defining the principle this way:

The principle of foudroyance has as its goal not destroying everything, which is without interest in any conflict, but breaking the rhythm or rhythms of the Other in its diverse activities, in such a way as to keep it from pulling itself together and to keep it a step behind the action.

This is exactly the goal of such information shock operations.

The potential for such paralytic information operations could fuel wishful thinking by contributing to an aggressor's overconfidence in its ability to win quickly and cheaply—a misperception which, as we explained in Chapter 3, is one of the most common routes to deterrence failure. It could offer new avenues to conducting attacks on an adversary's homeland (even if virtually) from the very first moments of a war, and thus undermine escalation control and potentially the stability of strategic balances. Our analysis thus reemphasizes the importance of securing U.S. national security and societal assets against such attacks.

## Lens 2: Effects on Deterrence Stability

Strategic stability describes a situation in which an actor is confident that its adversary (or potential adversary) is unable to undermine its deterrent capability; thus, both actors are in effect deterred from taking action against each other. Lens 2 explores the second major focus of our research—the stability of deterrent relationships, or the circumstances in which a current or emerging technology could threaten the stability of deterrent relationships by creating incentives to strike first because of a perceived instability of the deterrent situation. This analysis supported several key findings, summarized in this report and in Table 5.2.

**Table 5.2. Potential Effects on Deterrence Stability**

| Category of Effects | Risk to Stability of Deterrent Relationships | Technologies That Promise or Threaten Such Effects |
|---|---|---|
| Expanded scope of attack vectors | Increases uncertainty and fear and tempts states into unstable first strike or launch under attack options | • Cyber/information warfare<br>• Hypersonic systems<br>• Unmanned/swarming systems<br>• Biotechnology |
| Increased pace of conflict | Rapid pace of decisionmaking in crisis of war reduces space for deliberation, eliminates warning time, and impels faster responses with less information | • Cyber/information warfare<br>• DSSs<br>• Hypersonic systems |
| Arms racing | New technologies spur action-reaction cycle, which heightens underlying fears and uncertainties | • Hypersonic systems<br>• Unmanned systems<br>• General information categories |
| Ambiguity about use | Doubt about state of technology, how it will be employed, and degree of impact creates instability through uncertainty, act-first doctrines | • Biotechnology<br>• Quantum computing/technologies<br>• DSSs |
| Gray-zone escalation | Technologies can empower possibly escalatory gray-zone campaigns with false confidence in success | • Information and perception manipulation<br>• Cyber/information warfare<br>• Unmanned/swarming systems |
| Proliferation of technologies | Technologies that become accessible to non-state actors threaten existing deterrent relationships | • Biotechnology<br>• Cyber warfare and EW<br>• Information and perception manipulation<br>• Semiautonomous robotic systems |

First, our analysis suggested that new technologies can negatively affect stability by introducing new attack vectors or making current ones available to new adversaries, thereby increasing the reach and likelihood of hostile actions. The creation of new attack vectors could erode stability by causing an escalatory tit-for-tat interaction between adversaries, especially if these attacks are difficult to attribute and are perceived as falling below an escalation threshold. Uncertainty surrounding the use and intent of these technologies—particularly by nonstate actors, whose preferences are often less well-known—could further magnify effects on deterrence stability.

It may also become more difficult to predict or comprehend an adversary's decisionmaking process during a conflict if a new technology increases the pace of decisionmaking required for a response. Technologies, or combinations of technologies that impact decisionmaking timelines, will increase uncertainty surrounding those decisions, especially during a conflict. The uncertainty that accompanies an increased pace of decisionmaking and response could either prompt rapid escalation—by encouraging one side to strike preemptively out of concern that an attack is imminent given the ability of the other side to rapidly deploy its technology or by prompting quick retaliation once a first strike from either side occurs—or strengthen deterrence if both parties believe they would not be able to react quickly enough to the decisions of the other side.

We also found that new technology could push the pace of further technology development and acquisition as adversaries seek to remain ahead of their competitors, thus elevating the risk of escalation. The rapidly developing pace of new technologies would be especially likely to have such unstable effects in the context of an arms race environment.[6] A state must constantly be anticipating its adversaries' next moves technologically, or it could quickly and easily fall behind; thus, even if a state does not see a particular technology as being vital to its defense capabilities in general, if an adversary starts developing it or looks set to begin development or acquisition, the defending state must also pursue these technologies or risk being at a disadvantage. As new technologies are fielded on both sides, uncertainty over how adversaries may perceive and employ these technologies increases, complicating decisionmaking. This could, in turn, threaten the stability of the status quo and increase the risk of escalation (whether accidental or intentional).

Ambiguity surrounding a new technology and how it might be used generates additional instability. Uncertainty heightens tensions and generates incentives for first strikes. New technologies exacerbate this dynamic when there is uncertainty about readiness level, adversary intent, multidomain interactions, and the identity of adversaries. In addition, we found that the situational context may matter more than the specific weapon system or technology on its own; it

---

[6] For a discussion of the Chinese, Russian, and U.S. "arms race" dynamic playing out over AI, see Adrian Pecotic, "Whoever Predicts the Future Will Win the AI Arms Race," *Foreign Policy*, March 5, 2019.

is entirely dependent on whether a state views the holder of the technology or weapon system as being likely to use it, and ambiguity here could thus be destabilizing.

Additionally, some new technologies that generate effects primarily in the gray zone may have less clear escalatory dynamics associated with their use and may be a more appealing option for adversaries looking to destabilize targets while not provoking escalation into armed conflict. The theater of hostile actions is increasingly shifting to the *gray zone*—the space below the threshold of armed conflict. Use of such technologies as information- and perception-manipulation technologies that primarily have effects in the gray zone can create strategic instability yet may not upend conventional deterrence (an ideal outcome for adversaries seeking to inflict damage while not triggering all-out war). Some of the technologies also make attribution difficult, which could further incentivize states to take aggressive action in situations in which they would otherwise be deterred.

Increasing accessibility of technology to additional actors adds further complexity and uncertainty to existing deterrent relationships. Nonstate actors are increasingly becoming players in international security; empowered by emerging technologies, they can suddenly cause significant damage to states with little to no skill or effort (or cost, because some of these technologies, such as those in the information- and perception-manipulation category, are inexpensive and becoming fairly accessible). In addition, adversaries or other actors who have classically been at a disadvantage can target more-powerful actors like the United States by developing technologies that hit those actors in "blind spots" or gaps—places where powerful actors do not yet have a given technology, do not have recourse to respond for normative or legal reasons, or have overlooked because they have not traditionally been an area of strategic importance (e.g., a particular geographic area where they lack access).

In assessing the effects of technologies on the stability of deterrence, we also considered how arms control agreements, regulations, norms, and other such arrangements could mitigate or otherwise limit their effect on deterrence. This analysis is complicated by the fact that some technologies involve dual-use or commercial capabilities that are difficult to regulate.

An example of this qualification is exhibited in the development of biotechnology and potential weapons using this technology. A fundamental characteristic of biotechnology is its dual-use nature: It can be used to save lives through improving medical treatment or to take lives by creating chemical agents.

Because it is often impossible to determine with fidelity whether a particular capability will be used for good or ill, it is very difficult to regulate the use of biotechnology. Moreover, arms control depends on the ability to count or see deployed capabilities, thus systems that are less tangible/physical in nature or require a small footprint to deploy will be less amenable to arms control. This would include capabilities that do not require observable testing.

Finally, even if the technology is subject to control or regulation through arms control agreements, the current unravelling of the U.S.-Russia arms control regime demonstrates the speed with which these agreements can fall apart amid geopolitical tensions.[7] Because arms control agreements between feuding parties may be undermined in this manner, they may not hold as much deterrent power as expected (a topic covered at length in the previous section on Lens 1, which focuses on the credibility of deterrent threats).

Our analysis of the effects on stability highlighted two specific technologies with significant risks. The most obvious is hypersonic systems. Because of their incredible speed, these weapons compress decision timelines in ways that reduce crisis stability. Through their speed and precision, these systems specifically threaten command, control, communications, computers, and intelligence (C4I) nodes, thus posing at least a theoretical risk of exactly the sort of decapitation strike that U.S. rivals worry about. Hypersonic systems could also blur the boundaries between levels of conflict: During a major crisis or an accelerating gray-zone confrontation, one or more sides may determine that a small, localized hypersonic strike could gain advantage without requiring the large-scale use of force in ways that current long-range systems could not do without nuclear use or the employment of very large numbers of strike assets. The hypersonic arms race that is already underway is also likely to intensify mutual threat perceptions with dangerous results in periods of crisis.[8]

At the same time, there are limits to the negative effects that hypersonic systems could have. The most important is cost: Although precise cost figures for hypersonic weapons are not available in open sources, it is doubtful whether, especially under the financial pressures of the post-pandemic, high-debt era of the next decade, any major power will have sufficient resources to deploy enough of them to threaten strategic stability. As long as hypersonic systems cannot threaten a first strike against stable nuclear deterrents—and there is little reason to believe that they will be able to—nuclear weapons provide a reassuring ultimate deterrent against such risks as large-scale decapitation strikes. Unlike such capabilities as cyber, moreover, hypersonics are beyond the technical and financial capacities of smaller powers, reducing the dangers of unstable actions by third parties. Despite these constraints, our analysis suggests that hypersonic systems threaten the stability of deterrent balances in important ways.

Our research also highlighted a second technology with significant potential to undermine the stability of deterrent balances: automated response systems, which combine decision support with AI and, in some cases, semiautonomous or autonomous weapons. Major powers—and large numbers of middle powers—may be increasingly tempted to use these as the perceived speed of engagements rises. Our research suggested that both Russia and China have strong military

---

[7] For more on this topic, see: Samuel Charap, Alice Lynch, John J. Drennan, Dara Massicot, and Giacomo Persi Paoli, *A New Approach to Conventional Arms Control in Europe: Addressing the Security Challenges of the 21st Century*, Santa Monica, Calif.: RAND Corporation, RR-4346, 2020.

[8] In fact, since we completed our main analysis, Russia made a public statement declaring that it "will perceive any ballistic missile launched at its territory as a nuclear attack that warrants a nuclear retaliation," which clearly reinforces the escalatory risks of using hypersonic weapons. See Vladimir Isachenkov, "Russia Warns It Will See Any Incoming Missile as Nuclear," Associated Press, August 7, 2020.

traditions of employing or considering partly automated decision systems; both of those militaries may feel more comfortable competing at that level than by developing militaries comprising agile, flexible leaders who are able to operate on their own initiative without higher-level guidance. They could be employed for attack detection through algorithmic analysis of mass data sets for strategic and tactical warning even before conflicts begin. They could also be employed to make decisions during tactical engagements or even in a larger strategic sense—from automated choices of targets for air and missile defense to semiautomatic responses to various strategic moves.

The risks of such systems are obvious. Working at a speed of decision processing exceeding human pace would require delegation of decisionmaking to machines, which would reduce oversight and open the possibility of the system running out of control—as, for example, automated stock trading has done on several occasions. Various forms of bias can be baked into the assumptions and models of such systems. The workings of machine-learning algorithmic models are often opaque; it is one thing not to be aware of the reasons why a website is recommending certain books or music and another thing entirely to hand over military decisionmaking to an algorithmic process whose foundations remain obscure to the human users.

It is not clear how effective fail-safe mechanisms could be integrated into such systems. They could create an intense fear of first-mover advantage and speed of unfolding conflict that would risk producing hair-trigger, lightning-speed, machine-driven, interconnected systems highly sensitive to initial moves.

## Lens 3: Implications of Competitor Views of Deterrence

Whether deterrence strategies will be successful hinges on the motivations of potential aggressors—motivations that are necessarily state-specific rather than universal.[9] This analysis therefore undertook actor-specific analyses regarding the relationship between technology and deterrence for both Russia and China. We examined U.S. security and deterrence objectives vis-à-vis the state in question. We then, in turn, explored Russian and Chinese objectives vis-à-vis the United States and asked how technologies might affect those objectives, perceptions about their security, and the overall stability of deterrence. We then assessed national-level views in Russia and China regarding the anticipated effect of technology on the strategic landscape and its stability, including potential misperception factors fueled by different technologies, constraints on their ability and willingness to employ these weapons, and whether specific technologies are perceived to have significant interaction effects. Finally, we folded these insights into two military planning scenarios for each potential adversary, demonstrating how the various technologies examined in this project could plausibly affect perceptions about the security environment and operational planning.

---

[9] For example, see Mazarr et al., 2018; Mazarr, 2018; Morgan et al., 2008.

The resulting assessments are summarized in Table 5.3. These assessments support several primary findings.

**Table 5.3. Implications of Emerging Technologies for Specific Deterrence Challenges**

| Deterrent Challenge | Current Status | Possible Implications of Emerging Technologies |
|---|---|---|
| Russian intervention in the Baltics | No urgent Russian need to intervene; strong NATO commitment in place | *Minimal / Moderate*<br>• Would not change aggressor interests, motives<br>• Nonkinetic technologies empower gray-zone harassment, intrusions |
| North Korean invasion of South Korea | North Korea lacks key capabilities to invade; U.S. commitment remains strong | *Minimal*<br>• North Korea incapable of deploying, operating highest tech items<br>• Would not change geopolitical calculus and risks to aggressor |
| Chinese invasion of Taiwan | Chinese long-term intent clear, has not ruled out force; no imminent urgency | *Potentially Significant*<br>• Could enhance Chinese advantage<br>• Could provide ability to paralyze Taiwanese defense and U.S. response<br>• Nonkinetic technologies empower gray-zone harassment, intrusions |
| Chinese seizure of South China Sea | Chinese intent clear; current approach committed to gradualism | *Potentially Significant*<br>• Could enhance China's advantage, paralyze U.S./local target responses<br>• Nonkinetic technologies empower gray-zone harassment, intrusions |

First, both countries assign a high priority to technological competition but, in a general sense, are agnostic of specific technologies. Our review of both countries' thinking on deterrence and the role of emerging technologies did not reveal a view that any single technology or small set could be magic bullets to undermine U.S. deterrent policies. One partial exception to this rule is the strong emphasis in Chinese military planning on the role of information warfare and "systems destruction warfare," which is viewed as a way to gain decisive advantage in the early phases of conflict. This finding reinforces the importance, outlined above in the discussion of Lens 1, of such information capabilities in terms of their effect on deterrence.

Second, the two main U.S. rivals have different levels of ambition in the pursuit of emerging technologies. Russia generally seeks "balance" in technological relationships, which leaves more room for asymmetries without instability. Russia is also constrained in budgets, so high-cost technologies are out of reach at scale. China, on the other hand, makes consistent reference to leadership or dominance in key technologies. In some low-barrier-to-entry technological areas, such as cyber, Russia still has or could gain a significant ability to employ the systems. But broadly speaking, this difference in ambitions—and the resources and technological sophistication required to realize them—helps to narrow the focus of our findings. It is those deterrent requirements involving China that are most likely to be affected by emerging technologies.[10]

Third, when looking at specific technologies, both Russia and China emphasize hypersonic systems, although as a risk or threat as much as an opportunity. Both have emphasized the military applications of AI, although, in general terms and in Russia's case, without the associated resources to gain an advantage. Chinese sources have also highlighted biotechnology as an area of rising importance. Both also stress the great threat posed by another area: Any technologies that threaten an autocratic state's domestic control or overall C4I are seen as crucial and highly destabilizing if used.

Both competitors potentially embrace a more dangerous level of automated systems. Russian military thinking remains, to some degree, guided by quantitative models of "correlation of forces" and resulting implications, a legacy of Cold War automated detection/response. The Chinese leadership's preference for technological solutions could lead it to accept high levels of automated decision support and autonomous robotic systems without appropriate safeguards for escalation risks.

Fourth and finally, we assessed the country-specific aspects of technologies' effects on deterrence by looking at specific potential U.S. military contingencies as presented in Table 5.3. As the table suggests, in two cases—the risk of Russian adventurism in the Baltics and of North Korean aggression against South Korea—emerging technologies are unlikely to overcome multiple other variables involved in the deterrent relationship. It is again in regard to scenarios involving China, in both Taiwan and the South China Sea, that Beijing's mastery of key technologies could have the most potential effect.

## Lens 4: Effects of Technology Combinations on Deterrence

As discussed in previous chapters, deterrence entails discouraging an action or event through instilling doubt or fear of the consequences. We explored in Lens 4 the circumstances in which combinations of emerging technologies might have potential effects on deterrence, looking at threats to both credibility and stability. This could occur, for example, when a combination of two or more technologies changes the pace and dynamics of escalation, introduces a new vulnerability, or alters the power dynamics between actors.

---

[10] Previous RAND work compared China and Russia as potential threats, concluding that China was a peer competitor of the United States, and Russia was not. This was in part because of China's economic power and its technological ambitions. The authors concluded that

> Russia can be contained, employing updated versions of defense, deterrence, information operations, and alliance relationships that held the Soviet Union at bay for half a century. China cannot be contained. Its military predominance in east Asia will grow over time, compelling the United States to accept greater costs and risks just to secure existing commitments. (James Dobbins, Howard J. Shatz, and Ali Wyne, *Russia Is a Rogue, Not a Peer; China Is a Peer, Not a Rogue: Different Challenges, Different Responses*, Santa Monica, Calif.: RAND Corporation, PE-310-A, 2019)

To assess this issue, the project team assessed both whether and how each technology could combine with another to have an effect on deterrence, with particular emphasis on identifying specific causal routes through which such combinations could affect deterrence. It is important to note that the eight selected technologies can combine in nearly limitless ways to affect deterrence, so there are far more possible permutations than the ones discussed below. These are merely the combinations we chose to highlight because they seem to fall into the category of most likely *and* most dangerous. We did not assess how the deployment of technologies affects the need for and utility of others at a later point in time. Each of these combinations was analyzed in isolation.

Related to the competitor-specific analysis in Lens 3, our assessment found that both China and Russia see the value in investing at the intersection of several technology domains to alter power dynamics. Both Russian and Chinese military sources have discussed ways in which various emerging technologies could combine to achieve effects greater than the sum of their parts. We incorporated these findings in our general assessment of the combination effects of the technologies.

This analysis employed three primary analytical approaches to evaluating the effects of technology combinations:

1. *Identify mutual critical dependencies and acceleration/mitigation effects among the technologies.* When technologies combine, mutual dependencies and acceleration/mitigation effects commonly arise. For example, many combinations of technologies are critically dependent on AI; specifically capabilities developed for DSSs may also be used for command and control of groups of UAS/UAVs. Mitigation effects may apply to technologies that decrease decisionmaking timelines when deployed, such as incorporating DSSs into military command and control. As this technology improves, parts of decisionmaking become automated, meaning that responses could happen more quickly than human cognition. If hypersonic systems are potentially destabilizing because of the reduction in target response time, DSSs could both meet and mitigate this destabilizing factor by facilitating quicker responses.

2. *Identify obvious combination packages in which several technologies, both newly emerging and existing, could have effects on deterrence out of proportion to individual technologies.* This is, in a sense, the core question of this lens analysis, and our analysis sought to identify combination packages illustrating hypothetical yet realistic scenarios in which these packages have undeniable effects on international security and deterrence.

3. *Outline possible revolutionary concepts of warfare and how packages of technologies would empower them.* Although there are many ways that packages of technologies would empower revolutionary concepts of warfare, here, we will highlight two, both of which involve autonomous systems. The first is *the ability to detect and impose decisive costs on large-scale military aggression*, which ties autonomous systems to DSSs. This would likely require persistent deployment of airborne ISR and strike assets at a significantly larger scale than is currently possible with remote-controlled UAVs. The second is *merging levels or domains of conflict from tactical through operational to strategic*, tying autonomous systems to cyber warfare and EW. Large groups of stealth,

EW-hardened UAVs capable of persistent incursions to perform surgical strikes deep into enemy territory would translate into an utter lack of safe zones.

Our findings pointed to several leading potential implications, summarized in Table 5.4.

**Table 5.4. Potential Effects of Technology Combinations**

| Technology Combination | Risk to Deterrence | Technologies Included |
| --- | --- | --- |
| Information-centric combination | System destruction/network attack that offers the potential to paralyze or severely constrain a defender's response | • Cyber warfare and EW<br>• Information and perception manipulation<br>• DSSs<br>• EW-enabled UAS<br>• Quantum (eventually) |
| Battlefield strike complexes | Gain rapid destruction of enemy's fielded forces at the outset of war; includes some element of information combination but focus is more on precision strike | • UAS, especially for ISR<br>• Hypersonics<br>• DSSs<br>• Advanced sensing<br>• Cyber warfare and EW |
| Strategic strike complexes | Gain rapid wartime advantage through strikes on enemy homeland from first moments of war; broad-based paralytic effects | • Hypersonic systems<br>• Long-range UAS<br>• Cyber and information warfare<br>• Information and perception manipulation<br>• Biological agents |

Our analysis suggested that combinations of emerging technologies can both increase the lethality of current attack vectors and introduce new ones, resulting in an increased likelihood of hostile actions. Mixes of emerging technologies can also provide an increased ability to reach new or existing targets, which can in turn increase the probability of a first strike. This can be true both in kinetic terms (e.g., using hypersonic systems to reach new targets) or virtual (e.g., using electronic or information warfare capabilities).

As we have stressed, these military effects—especially on the effectiveness of U.S. deterrent threats—depend substantially on the *balance* of these technologies between the two sides. Moreover, we sought to assess ways in which new technologies could enhance U.S. deterrent policies and those that would undermine it. Rapid mutual development and deployment of unmanned systems, for example, might create new military dynamics but not produce a perceived advantage on either side that would either erode or strengthen deterrence. Our analysis highlighted numerous areas in which U.S. competitors appear to be developing packages of technologies with the potential to achieve sufficient perceived advantage to be dangerous for deterrence effectiveness but also areas of opportunity for the United States to respond and enhance deterrent effects. We highlight those in Chapter 6.

Combinations of emerging technologies also hold the potential to increasingly blur boundaries in military operations. Several specific emerging technologies inherently break down such boundaries: Cyber, EW, and disinformation capabilities are employed below the threshold of major conflict but also to set the conditions for, and prevail in, military operations. Longer-range strike systems, including UAS and hypersonics, may weaken the barrier between close battle and deep maneuver, creating new potential for escalatory dynamics: If one combatant has what it perceives to be easier, more-effective non-nuclear means for striking at command and control targets in its opponent's homeland, this may accelerate the escalation of local fights to generalized war. Emerging technologies are also blurring the boundary between operational warfighting areas and home turf, providing non-nuclear means—both virtual and kinetic—of rapidly drawing homelands into conflicts.

Finally, our analysis of possible technology combinations and their effects on deterrence holds a broader lesson: Warfare is taking on a new character that we must operate within but do not fully understand. We will discuss this finding in more detail in the concluding chapter. Grappling with the implications of developing effects that emerging technology could have on deterrence—especially how those technologies combine into various packages to achieve larger effects—will require thinking in detail about new models of major combat and the conceptual foundations for them.

# 6. Findings and Conclusions

The previous chapter outlined potential effects of the eight selected technology areas examined in this report on the effectiveness and stability of deterrent relationships . We then sought to assess the broader implications of those findings—what they mean for future U.S. national security strategy and for the USAF in particular. This chapter lays out these broader implications, which are informed by our findings from the totality of research for the project (described in this summary report) and the specific effects on deterrence reviewed in Chapter 5.

Our research highlights two overarching conclusions. First, *collections of emerging technologies*—especially in the realms of information aggression and manipulation, automation (including automated DSSs), hypersonic systems, and unmanned systems—*hold dramatic implications for both the effectiveness and stability of deterrence*. These risks may call for changes in U.S. policies, operational concepts, and technology development programs. In some cases, they may point to the value of arms control or confidence-building regimes. The USAF and wider national security community should give these risks significant attention.

An important implication is that, properly conceived and implemented, such combinations of technologies could bolster the effectiveness of U.S. deterrent policies rather than undermine them. In its growing embrace of multidomain operations, joint all-domain command and control (JADC2), and related concepts that speak to a holistic integration of multiple instruments of military power, the USAF is already embracing approaches that speak to the integrated application of technologies. Our findings suggest that such combinations potentially provide the best avenues to strengthening deterrent threats, but the specific operational and perceptual effects of such combinations remain poorly understood. Multidomain operations and JADC2 themselves are still aspirational statements of intent more than fully developed action plans. Our analysis suggests that future thinking on deterrence should focus on such technology mixes and how they can best affect potential aggressor perceptions.

Second, *an emerging transition to new ways of warfare, empowered by these same emerging technologies, poses more general risks to U.S. deterrent policies than any single technology or set of them*. The effects of technologies on deterrence are an important but only partial subset of a more profound reality—the changing character of warfare toward more information-based, unmanned, semiautonomous, and AI-driven platforms. If the United States is left behind in the technological but also conceptual and doctrinal transition to this new era, both the effectiveness and stability of U.S. deterrent policies are likely to suffer.

## General Findings

The analysis of the preceding chapters supports a number of key general findings about the relationship between emerging technologies and deterrence.

*With some exceptions, the effects of individual technologies or military systems are likely to remain an enabler, not a prime cause, of deterrence failure.* Improved capabilities at margins are rarely if ever the decisive factor in deterrence failure.[11] Relative advances in one or two specific technologies can make some difference in war outcomes and thus potentially threaten deterrence. But alone they are seldom the decisive factor in deterrence failure. War is a political act, not the result of immutable technological forces; without the necessary motives, states will not risk major war, especially in the nuclear era, merely because of a momentary window of opportunity afforded by technological advances. More-serious dangers arise when powerful geopolitical ambition is married to a highly effective military empowered by an innovative operational concept, which is itself fueled by mutually supporting technological advances. The single case in which it could be true today—although the analogy is imperfect—is China's use of an information warfare/system-destruction warfare concept and associated technologies in service of its regional ambitions, especially involving Taiwan.

*The strategies that military organizations use to employ technologies are critical to understanding their effects on deterrence.* In the same way that individual technologies do not decisively affect military outcomes, the effects they do have are shaped by an important intervening variable—the choices that military organizations make on how to employ them. These include, most critically, the operational concepts developed to make packages of capabilities work together to achieve outcomes. This fact reinforces one of the fundamental conclusions of this report: that the United States should focus on net assessment of competitor concepts, and development of U.S. concepts, that embody such synergistic theories of success.

*As a result, the risks of deterrence failure are greatest in scenarios in which multiple technologies work together to exacerbate classic sources of deterrence failure.* Individual systems rarely have a transformational effect on their own, and combinations of technologies can fuel a seemingly decisive operational concept to encourage wishful thinking and belief in the ability to achieve desired gains in a short, relatively low-cost war.

*No technology appears to threaten the effectiveness of core U.S. nuclear deterrence policy* (although stability may be a different consideration). None of the technologies we examined appears to have the ability to disable the U.S. nuclear deterrent. Two technologies—information warfare and hypersonic weapons—are partial exceptions to this rule because each could be used as part of a broader effort to disable the U.S. nuclear arsenal. But, given present constraints,

---

[11] Two leading studies that assessed the role of technological advance in warfare outcomes came to similar conclusions—that combinations of technologies, linked to a coherent operating or force employment concept and adopted by highly effective militaries, can make a decisive difference, but that individual technologies alone seldom do. See Lieber, 2008, and Biddle, 2006.

these two technologies are unlikely to pose a fundamental threat to the U.S. nuclear deterrent. Nor does open-source evidence suggest that U.S. rivals believe that they can transcend nuclear deterrence using those or other technologies.

*Technology combinations complicate deterrence by offering the potential to hit multiple targets across many attack surfaces simultaneously.* Rather than merely attacking fielded military forces, an attacker could soon use information warfare (including attacks on space-based assets), hypersonic systems, UAVs (including some with intercontinental range), AI-driven DSSs, and other technologies to launch simultaneous blows against a defender's entire military, economic, governmental, and social systems. This creates an opportunity for society-wide paralytic attacks that could undermine deterrence by providing an aggressor with the possible false hope of being able to freeze the defender long enough to achieve desired gains. This is one area in which a handful of technologies, employed in service of a coherent operating concept, could threaten the effectiveness of deterrence in fundamental ways. But it also points to potential opportunities for the United States to pursue its own versions of such combinations to enhance deterrent policies.

*In terms of contingencies of major warfare, technologies have the greatest potential to degrade the effectiveness of deterrence in scenarios involving China.* Even when used in combination packages, and even assuming their greatest potential effects, the technologies surveyed in this analysis are not likely to shift the effectiveness of deterrent relationships in Europe or Korea. This is true for at least four reasons.

First, the European and Korean deterrence situations are either overdetermined or decisively affected by other factors. We did not assess the North Korea threat in detail, but Pyongyang's nuclear arsenal, for example, is likely to have a much greater effect on deterrence than any new technologies. China's willingness to employ force in service of its claims in regard to Taiwan and the South China Sea appears to be greater than current Russian or North Korean willingness to use force over the Baltics or reunifying the Peninsula.

Second, neither Russia nor North Korea are making anything like the investments in these new technologies that China is making. Neither has a technological base to permit widespread acquisition and use of more than one or two of the technologies. China is a unique technological-industrial challenge in this regard. Third, this analysis highlighted the fact that China, more than any other U.S. challenger, has a declared intention to use combinations of emerging technologies for decisive military effects. Fourth and finally, the geography and geopolitics of the scenarios suggest that some of the deterrence effectiveness effects discussed here will apply to China more than other potential aggressors: South Korea and the Baltics are formal U.S. allies with U.S. (and in the case of Europe, NATO) ground forces on their territory. The Indo-Pacific scenarios offer much more room for Beijing to use technology packages to convince itself that it can engage in a *fait accompli* while paralyzing the U.S. response.

Using these considerations, we conclude that, by far, the most likely scenarios in which technology poses a true threat to the effectiveness of U.S. deterrent threats are two contingencies involving China: Taiwan and the South China Sea.

*Multiple interacting forms of automation carry very significant risks, especially for the stability of deterrent relationships.* One of the most serious hazards emerges from automation. Over the next two decades, the stability of strategic balances could be upset by the interaction of multiple AI-driven automated systems in a fundamentally new form of warfare. In an extreme circumstance, there may be dangers of elaborate mechanisms that escalate very quickly from local engagements to global war, driven by automated systems. U.S. competitors appear to have greater tolerance for the risks of such systems and may come to rely on them in major ways.

*Non–AI-based technologies also threaten the stability of deterrent relationships.* Beyond automated systems, including AI-driven decision support, other specific technologies we examined (including cyber weapons, hypersonic systems, biotechnology-enabled disruption, and robotic systems) carry potential destabilizing effects by increasing the premium on striking first in a crisis and reducing each side's confidence in its ability to withstand attacks.

*Many technologies challenge the U.S. ability to deter aggression, coercion, and influence-seeking below the threshold of major war in the "competition phase."* In addition to their relevance to high-end conflict, nearly all the technologies we examined have potential uses in enabling coercive acts below the threshold of war. Cyber capabilities, disinformation, unmanned systems, biological tools, and even AI-driven DSSs could strengthen and increase the frequency of bellicose actions in the gray zone.

*There is a growing potential for information- and perception-manipulation technologies, including deepfakes, to contribute to the failure of deterrence.* Aggressors are likely to make growing use of information- and perception-manipulation technologies and techniques to create confusion, delay responses, divide alliances, and achieve other effects that could encourage wishful thinking or promote strategic miscalculations about war outcomes and thus undermine deterrence effectiveness.

*On the opportunity side of the ledger, the United States could employ emerging technologies to enhance the effectiveness and stability of deterrence and stability in multiple ways.* These include

- investments to gain advantage in the contest for resilience against systemic attack and counter–systems warfare in the information space
- the use of UAVs/UAS, AI-driven analysis, and cyber capabilities as part of a network of persistent, comprehensive domain awareness and targeting capabilities to enhance awareness of and transparency in gray-zone activities, warning of large-scale military operations, and verification of rules of engagement and agreements or norms on the limitation of military activities
- networks of new-generation precision-guided weapons married to UAS and DSSs to enhance targeting and intensify the threat to any advancing forces

- the transfer of technology, including co-development, to allies and partners to enhance their independent capability to deter and defeat aggression.

In sum, the technologies surveyed in this analysis carry the potential to significantly degrade the effectiveness of deterrence, but this potential is largely confined to two very specific scenarios (Chinese intervention in Taiwan and the South China Sea) and specific packages of technologies focused on disrupting information and communication networks. These technologies pose more-general threats to the stability of deterrence. In both cases, these effects demand more debate and study. Broadly, these technologies have the potential to usher in a new era of deterrence and warfighting that could either benefit the United States or pose new risks to its deterrent policies.

## Implications for the U.S. Air Force

These findings, as well as the broader analysis conducted for this report, hold several implications for the USAF.

- *To remain attuned to deterrence risks, focus on understanding the perceptions of rivals first and the technology second.* Especially for a service like the Air Force, with its strong institutional and cultural affinity for technology, it could be tempting to think of the problem of deterrence as primarily technical. In fact, deterrence is a political, not a technical, challenge; reasons for deterrence failure typically have far more to do with the views and ambitions of potential aggressors than specific technologies. The foundation for ongoing thinking about deterrence is therefore a deep and consistent appreciation for current Russian and Chinese decision theory, perceptions of possible contingencies, the role of technology in them, and the possible routes to wishful thinking—or fear—in threatening deterrence.
- *The USAF should place special emphasis on awareness of the technology packages that near-peer adversaries are investing in and how they seek to combine them.* One implication is to demand a joint and all-domain mindset in thinking about technology threats to deterrent capabilities. Wargames and exercises could test the potential effect of such technology packages on U.S. response and warfighting capabilities. This same focus calls for attention to competitor perceptions of U.S. technology combinations—both for assessments of successful deterrent signaling (do they find the combinations credible and daunting?) and risks of provocation (do specific technology combinations create a fear of surprise attack?).
- *As part of the continuing development of such concepts as multidomain operations and JADC2, the USAF should assess both the operational effects of such combinations and the effects on the perceptions of potential aggressors.* Combinations or synergistic mixes of emerging technologies offer opportunities as well as risks. Notwithstanding the conceptual development that has already occurred in cross-domain applications of technologies, there has still been relatively little analysis of the detailed operational effects of various combination packages or their possible effect on aggressor perceptions. Such analysis could underwrite the development of technology packages with important value for deterrence and should be a priority for the USAF.

- *Securing against information network/Chinese system-destruction attacks is a precondition for effective deterrence and stability.* Dominance at the level of exquisite systems and munitions will not be enough when potential aggressors hope to void U.S. advantages in those areas by fracturing U.S. information networks. Despite the existence of many cyber offices, activities, and initiatives, the USAF's degree of preparedness to deal with this threat remains unclear. Conducting a highly demanding, service-wide "network resilience audit" would be an important first step.

- *The UAS/counter-UAS competition is likely to become a major focus of U.S. defense investments and the stability of deterrent relationships in key theaters.* As in the case of network security, the USAF would benefit from a general audit of where it stands relative to major competitors in this area—especially because the competition spans multiple domains and services and risks becoming fragmented across dozens of poorly coordinated programs. The USAF could also develop enhanced concepts and proposed capabilities for ways in which such systems could enhance domain awareness in gray-zone campaigns, including in open-source information.

- *Norms, rules, and limits governing technologies could benefit the United States.* The risks embodied in a number of these technologies, particularly to stability but also to deterrence effectiveness, could be mitigated with confidence-building and arms-limitation agreements, as well as more-general norms of conduct that represent sufficient degrees of shared interests that they are likely to be at least substantially respected by others. These include rules of engagement for UAS operations; upgraded crisis-management tools in general, including crisis communication links; limits on the deployment of fully autonomous weapon systems; and agreed norms governing both homeland cyberattacks and the use of automated decision systems. Another such limitation that could benefit the United States is a limit on the number of allowed hypersonic systems: Besides their risk to stability, their significant cost will increasingly present trade-offs against other USAF priorities, and it is not likely to be an area where the United States can achieve a decisive advantage.

- *Building relationships with rival air force leaders can provide important benefits.* Many of these technologies carry the risk of dramatically accelerating the pace of conflict and injecting new sources of escalatory instability into crises and wars. These trends increase the value of relationships; if senior USAF leaders can cultivate personal ties with senior counterparts in Russia and China, the relationships can then be leveraged to help reduce the risk of miscalculation or misperception.

- *Technology integration in support of concepts of warfare will be increasingly crucial.* Because combinations of these technologies will be especially impactful for both deterrence and broader warfighting effects, the effectiveness of the USAF will depend to an ever-increasing degree on its ability to integrate capabilities from disparate technology areas (which is to be expected in the era of all-domain operations). Stove-piping of capability and concept development, and systems that do not adequately mesh together, will be more dangerous to USAF operations than ever. But it is not clear who owns

USAF technology-integration responsibility—and the parallel responsibility for fully integrating with joint capabilities outside the USAF—especially at the conceptual level.[12]

- *Given the effects on deterrence relationships adduced in this report, the United States could benefit from expanded multilateral development of systems likely to offer significant deterrent value—including sensing systems, UAS, and precision weapons—for partner or ally self-defense.* Limits to U.S. power projection capabilities, which are emerging partly because of the military effects of some of the very technologies examined in this report (such as cyber, hypersonic systems, and semiautonomous systems), will require an increase in U.S. reliance on allies and partners to fight immediate and even long-term local battles largely on their own. Working in collaboration with allies and partners to develop technologies to create an ability for a defender to destroy attacking forces will be a particular priority.

- *Finally, the USAF should also assess the potential institutional effects—positive as well as negative—of these emerging technologies.* Apart from the military operational implications of these technologies, they also carry possible opportunities and risks. Information- and perception-manipulation capabilities, for example, could be used to undermine USAF recruiting campaigns or to target specific USAF senior leaders for campaigns of harassment. The rise of a UAS-centric force will have implications for service culture, careers, recruiting, and retention. Many of these technologies will affect the USAF as an institution as much or more than they affect deterrence as a practice, and the USAF would be well served to anticipate such possible outcomes.

## Technology, Deterrence, and New Ways of Warfare

Finally, the research articulated in this report offers a lesson that goes beyond the effects of individual technologies on deterrence and speaks to the broader issue of the future of warfare. Although the true character of the emerging form of war remains to be determined, some very likely elements now seem clear and were reflected in many of the technologies we assessed. Future wars among major powers are likely to feature (among other characteristics) an intense effort to destroy an adversary's information acquisition, processing, networking, and assessment capabilities (both terrestrial and space-based, both within the theater and beyond); much greater use of unmanned and semiautonomous systems; increased reliance on dispersal and concealment to enhance survivability; and gradual integration of AI-powered DSSs. Together, the effects of these trends reinforce the emphasis on all-domain warfare and call on the USAF to intensify its efforts to build concepts appropriate for the emerging era.

One implication of these trends is to raise the possibility of truly "boundaryless" warfare with tactical, operational, strategic, and homeland targets, military as well as civilian, all engaged at the same time from the first moments of warfare. The combined effect of these technologies is to

---

[12] This is not the same as USAF-wide technology development or R&D strategy, which are now coordinated under a chief technical officer. The argument here is in terms of the operational warfighting (and competition-phase) effects of various technologies and how they nest together.

reduce significance of range, blur boundaries between levels of war and the battlefield and homeland, and fundamentally expand the attack surface.

Warfare in the past has had some of these characteristics—civilian and homeland targets certainly featured prominently in World War II—but these 21st-century technologies have the potential to radically expand these realities and engage an unprecedented range of targets in conflict. These targets could range from USAF bases in the theater of operations to American perceptions about the personal safety and security of specific USAF unit commanders to the coherence of U.S. domestic information networks.

Among other things, these trends demand new thinking about concepts of operations and how the USAF will fight in such an environment. This research suggests that the notion of JADC2 as currently defined represents only a small part of this larger future. It is an important step forward, but only the tip of the iceberg, in recognizing the emerging all-domain reality. Among other forms of competition, the United States will be engaged in a contest to define and implement force-employment concepts that use emerging technologies to achieve competitive advantage.

History suggests that, within the military sphere, this conceptual space may harbor the most-pivotal areas of competition. Transformative operating concepts that make best use of new technologies and are effectively implemented by high-quality militaries have repeatedly provided decisive military advantage in wartime. Military analyst Steven Biddle, for example, has catalogued the emergence of the post-1918 combined arms or "modern system" approach to force employment, which allowed selected militaries to overcome the growing lethality of firepower with "cover, concealment, dispersion, [and] suppression," small-unit maneuver, and the use of reserves and means of fighting in depth. "Militaries that fail to implement the modern system," Biddle concluded, "have been fully exposed to the firepower of modern weapons—with increasingly severe consequences."[13]

Multiple combinations of emerging technologies suggest the potential of moving toward a new approach to force employment, the effect of which will be at least as transformative as the modern system described by Biddle. These concepts will be grounded in information-network destruction and manipulation and point toward a future characterized in part by unmanned, semiautonomous, dispersed, and swarming systems. The major powers that master this new approach will have a decisive advantage in war and possess predominant power and influence. Getting the understanding of conceptual models for force employment wrong carries significant risk, both in terms of the effectiveness of U.S. deterrent threats and the potential to prevail in major war.

This study's final implication, therefore, is that the development—and, just as important, effective implementation—of force-employment concepts appropriate to the capabilities of emerging technologies is arguably the single most important priority for the USAF. This process

---

[13] Biddle, 2006, p. 3.

is well underway, with the development of such notions as JADC2 and the work of the relatively new Air Force Warfighting Integrating Capability, particularly its Futures and Concepts Division. At the joint level, work on future joint force concepts under the broad all-domain operations rubric continues.

This report highlights the critical imperative of adequately resourcing and pushing this conceptual development process. But it also points to the potential risk that these efforts would not produce concepts that are migrated to the force and fully implemented, so that the USAF and joint force are prepared to train, exercise, and posture to accomplish them in time for a major contingency. Apart from their direct effects on deterrence, therefore, the emerging technologies assessed for this analysis emphasize the critical importance of not being left behind in the race to master the force employment concepts of the new era of warfare.

# Abbreviations

| | |
|---|---|
| 5G | fifth generation |
| AI | artificial intelligence |
| C4I | command, control, communications, computers, and intelligence |
| DARPA | Defense Advanced Research Projects Agency |
| DEW | directed-energy weapon |
| DoD | U.S. Department of Defense |
| DSTL | Developing Science and Technologies List |
| DSS | decision support system |
| EM | electromagnetic |
| EW | electronic warfare |
| GPS | Global Positioning System |
| HCM | hypersonic cruise missile |
| HGV | hypersonic glide vehicle |
| IDSS | intelligent decision support system |
| IoT | Internet of Things |
| ISR | intelligence, surveillance, and reconnaissance |
| JADC2 | joint all-domain command and control |
| LAWS | lethal autonomous weapon systems |
| MAC | media access control |
| MCTL | Militarily Critical Technologies List |
| NSS | National Security Strategy |
| R&D | research and development |
| UAS | unmanned aircraft system |
| UAV | unmanned aerial vehicle |
| USAF | U.S. Air Force |

# References

Alexandroff, Alan, and Richard Rosecrance, "Deterrence in 1939," *World Politics*, Vol. 29, No. 4, April 1977, pp. 404–424.

Altmann, Jürgen, and Frank Sauer, "Autonomous Weapon Systems and Strategic Stability," *Survival*, Vol. 59, No. 5, 2017, pp. 117–142.

Arquilla, John, and David Ronfeldt, *Swarming and the Future of Conflict*, Santa Monica, Calif.: RAND Corporation, DB-311-OSD, 2000. As of July 20, 2020:
https://www.rand.org/pubs/documented_briefings/DB311.html

Biddle, Stephen, *Military Power: Explaining Victory and Defeat in Modern Battle*, Princeton, N.J.: Princeton University Press, 2006.

Bidwell, Christopher A., and Bruce W. MacDonald, *Special Report: Emerging Disruptive Technologies and Their Potential Threat to Strategic Stability and National Security*, Washington, D.C.: Federation of American Scientists, September 2018. As of June 12, 2021:
https://fas.org/wp-content/uploads/media/FAS-Emerging-Technologies-Report.pdf

Brose, Christian, *The Kill Chain: Defending America in the Future of High-Tech Warfare*, New York: Hachette Books, April 21, 2020.

Buchholz, Scott, *Tech Trends 2019: Government and Public Services Perspective*, Deloitte Consulting LLP, 2019. As of June 12, 2021:
https://www2.deloitte.com/content/dam/Deloitte/us/Documents/public-sector/us-gps-government-tech-trends-2019.pdf

Buchholz, Scott, Joe Mariani, Adam Routh, Akash Keyla, and Pankaj Kamleshkumar Kishnani, "The Realist's Guide to Quantum Technology and National Security," *Deloitte Insights*, February 6, 2020. As of July 5, 2021:
https://www2.deloitte.com/us/en/insights/industry/public-sector/the-impact-of-quantum-technology-on-national-security.html

Charap, Samuel, Alice Lynch, John J. Drennan, Dara Massicot, and Giacomo Persi Paoli, *A New Approach to Conventional Arms Control in Europe: Addressing the Security Challenges of the 21st Century*, Santa Monica, Calif.: RAND Corporation, RR-4346, 2020. As of August 18, 2020:
https://www.rand.org/pubs/research_reports/RR4346.html

Crane, Keith W., Lance G. Joneckis, Hannah Acheson-Field, Iain D. Boyd, Benjamin A. Corbin, Xueying Han, and Robert N. Rozansky, *Assessment of the Future Economic Impact of Quantum Information Science*, Washington, D.C.: IDA Science and Technology Institute,

August 2017. As of July 5, 2021:
https://www.ida.org/-/media/feature/publications/a/as/assessment-of-the-future-economic-impact-of-quantum-information-science/p-8567.ashx

Davis, Zachary, "Artificial Intelligence on the Battlefield: Implications for Deterrence and Surprise," *Prism*, Vol. 8, No. 2, 2019, pp. 115–131.

Defense Innovation Unit, *Annual Report 2018*, 2018. As of August 31, 2020:
https://assets.ctfassets.net/3nanhbfkr0pc/26OECgBCK7AZ2XhkbLyyMQ/3a64bb032550e3f0b9086b00d60f012c/DIU_2018_Annual_Report_FINAL.pdf

Department of Defense Directive 3000.09, *Autonomy in Weapon Systems*, Washington, D.C., incorporating change 1, May 8, 2017. As of July 9, 2021:
https://www.esd.whs.mil/portals/54/documents/dd/issuances/dodd/300009p.pdf

Dobbins, James, Howard J. Shatz, and Ali Wyne, *Russia Is a Rogue, Not a Peer; China Is a Peer, Not a Rogue: Different Challenges, Different Responses*, Santa Monica, Calif.: RAND Corporation, PE-310-A, 2019. As of July 9, 2021:
https://www.rand.org/pubs/perspectives/PE310.html

Dobber, Tom, Ronald Ó. Fathaig, and Frederik J. Zuiderveen Borgesius, "The Regulation of Online Political Micro-Targeting in Europe," *Internet Policy Review*, Vol. 8, No. 4, December 2019, pp. 1–20.

DoD—*See* U.S. Department of Defense.

Fári, M. G., and U. P. Kralovánsky, "The Founding Father of Biotechnology: Károly (Karl) Ereky," *International Journal of Horticultural Science,* Vol. 12, No. 1, 2006, pp. 9–12.

Federal Communications Commission, "Caller ID Spoofing," webpage, March 7, 2021. As of June 12, 2020:
https://www.fcc.gov/consumers/guides/spoofing-and-caller-id

Foote, Keith D., "A Brief History of Machine Learning," Dataversity webpage, March 26, 2019. As of March 5, 2020:
https://www.dataversity.net/a-brief-history-of-machine-learning/

Gartzke, Erik, "Blood and Robots: How Remotely Piloted Vehicles and Related Technologies Affect the Politics of Violence," *Journal of Strategic Studies*, October 3, 2019, pp. 764–788.

Geist, Edward, and Andrew J. Lohn, *How Might Artificial Intelligence Affect the Risk of Nuclear War?* Santa Monica, Calif.: RAND Corporation, PE-296-RC, 2018. As of August 18, 2020:
https://www.rand.org/pubs/perspectives/PE296.html

Grissom, Adam, "The Future of Military Innovation Studies," *Journal of Strategic Studies*, Vol. 29, No. 5, 2006, pp. 905–934.

Hopf, Ted, *Peripheral Visions: Deterrence Theory and American Foreign Policy in the Third World, 1965–1990*, Ann Arbor, Mich.: University of Michigan Press, 1994.

Horowitz, Michael C., *The Diffusion of Military Power: Causes and Consequences for International Politics*, Princeton, N.J.: Princeton University Press, 2010.

Horowitz, Michael C., "When Speed Kills: Lethal Autonomous Weapon Systems, Deterrence and Stability," *Journal of Strategic Studies*, Vol. 42, No. 6, 2019, pp. 764–788.

Horowitz, Michael C., "Do Emerging Military Technologies Matter for International Politics?" *Annual Review of Political Science*, Vol. 23, May 2020, pp. 385–400.

Huth, Paul K., *Extended Deterrence and the Prevention of War*, New Haven, Conn.: Yale University Press, 1988.

Huth, Paul K., "Deterrence and International Conflict: Empirical Findings and Theoretical Debates," *Annual Review of Political Science*, Vol. 2, No. 1, June 1999, pp. 25–48.

Isachenkov, Vladimir, "Russia Warns It Will See Any Incoming Missile as Nuclear," Associated Press, August 7, 2020. As of August 18, 2020:
https://apnews.com/888e0816c6fa7f58b9ad4f1e97993643

Jervis, Robert, "Deterrence and Perception," *International Security*, Vol. 7, No. 3, Winter 1982–1983, pp. 3–30.

Joint Publication 3-0, *Joint Operations*, Washington, D.C.: Joint Chiefs of Staff, January 17, 2017, incorporating change 1, October 22, 2018. As of June 21, 2021:
https://www.jcs.mil/Portals/36/Documents/Doctrine/pubs/jp3_0ch1.pdf

Joint Publication 3-12, *Cyberspace Operations*, Washington, D.C.: Joint Chiefs of Staff, June 8, 2018. As of June 21, 2021:
https://www.jcs.mil/Portals/36/Documents/Doctrine/pubs/jp3_12.pdf

Joint Publication 3-85, *Joint Electromagnetic Spectrum Operations*, Washington, D.C.: Joint Chiefs of Staff, May 22, 2020. As of June 21, 2021:
https://www.jcs.mil/Portals/36/Documents/Doctrine/pubs/jp3_85.pdf

Kinsella, David, and Bruce Russett, "Conflict Emergence and Escalation in Interactive International Dyads," *Journal of Politics*, Vol. 64, No. 4, November 2002, pp. 1045–1068.

Lebow, Richard Ned, "The Deterrence Deadlock: Is There a Way Out?" *Political Psychology*, Vol. 4, No. 2, June 1983, pp. 333–354.

Lebow, Richard Ned, "Windows of Opportunity: Do States Jump Through Them?" *International Security*, Vol. 9, No. 1, Summer 1984, pp. 147–186.

Lebow, Richard Ned, and Janice Gross Stein, "Deterrence: The Elusive Dependent Variable," *World Politics*, Vol. 42, No. 3, April 1990, pp. 336–369.

Leng, Russell J., "Escalation: Competing Perspectives and Empirical Evidence," *International Studies Review*, Vol. 6, No. 4, 2004, pp. 51–64.

Libicki, Martin C., *Cyberdeterrence and Cyberwar*, Santa Monica, Calif.: RAND Corporation, MG-877-AF, 2009. As of August 18, 2020:
https://www.rand.org/pubs/monographs/MG877.html

Lieber, Keir A., *War and the Engineers: The Primacy of Politics over Technology*, Ithaca, N.Y.: Cornell University Press, 2005.

Lieber, Keir A., "The New Era of Counterforce: Technological Change and the Future of Nuclear Deterrence," *International Security*, Vol. 41, No. 4, Spring 2017, pp. 9–49.

Mattis, Jim, *Summary of the 2018 National Defense Strategy: Sharpening the American Military's Competitive Edge*, Washington, D.C.: U.S. Department of Defense, 2018. As of June 21, 2021:
https://dod.defense.gov/Portals/1/Documents/pubs/2018-National-Defense-Strategy-Summary.pdf

Mazarr, Michael J., *Understanding Deterrence*, Santa Monica, Calif.: RAND Corporation, PE-295-RC, 2018. As of July 20, 2020:
https://www.rand.org/pubs/perspectives/PE295.html

Mazarr, Michael J., Ryan Michael Bauer, Abigail Casey, Sarah Heintz, and Luke J. Matthews, *The Emerging Risk of Virtual Societal Warfare: Social Manipulation in a Changing Information Environment*, Santa Monica, Calif.: RAND Corporation, RR-2714-OSD, 2019. As of July 20, 2020:
https://www.rand.org/pubs/research_reports/RR2714.html

Mazarr, Michael J., Arthur Chan, Alyssa Demus, Bryan Frederick, Alireza Nader, Stephanie Pezard, Julia A. Thompson, and Elina Treyger, *What Deters and Why: Exploring Requirements for Effective Deterrence of Interstate Aggression*, Santa Monica, Calif.: RAND Corporation, RR-2451-A, 2018. As of July 20, 2020:
https://www.rand.org/pubs/research_reports/RR2451.html

McFate, Sean, *The New Rules of War: Victory in the Age of Durable Disorder*, New York: William Morrow, 2018.

Mearsheimer, John J., "The German Decision to Attack in the West, 1939–1940," in *Conventional Deterrence*, Ithaca, N.Y.: Cornell University Press, 1983, pp. 99–133.

Ménard, Alexandre, Ivan Ostojic, and Mark Patel, "A Game Plan for Quantum Computing," *McKinsey Quarterly*, February 6, 2020. As of July 5, 2021:
https://www.mckinsey.com/business-functions/mckinsey-digital/our-insights/a-game-plan-for-quantum-computing

Mercer, Jonathan, *Reputation and International Politics*, Ithaca, N.Y.: Cornell University Press, 1996.

Morgan, Forrest E., *Deterrence and First-Strike Stability in Space: A Preliminary Assessment*, Santa Monica, Calif.: RAND Corporation, MG-916-AF, 2010. As of August 18, 2020: https://www.rand.org/pubs/monographs/MG916.html

Morgan, Forrest E., Karl P. Mueller, Evan S. Medeiros, Kevin L. Pollpeter, and Roger Cliff, *Dangerous Thresholds: Managing Escalation in the 21st Century*, Santa Monica, Calif.: RAND Corporation, MG-614-AF, 2008. As of August 18, 2020: https://www.rand.org/pubs/monographs/MG614.html

Morris, Lyle J., Michael J. Mazarr, Jeffrey W. Hornung, Stephanie Pezard, Anika Binnendijk, and Marta Kepe, *Gaining Competitive Advantage in the Gray Zone: Response Options for Coercive Aggression Below the Threshold of Major War*, Santa Monica, Calif.: RAND Corporation, RR-2942-OSD, 2019. As of June 21, 2021: https://www.rand.org/pubs/research_reports/RR2942.html

Mueller, Karl P., Jasen J. Castillo, Forrest E. Morgan, Negeen Pegahi, and Brian Rosen, *Striking First: Preemptive and Preventive Attack in U.S. National Security Policy*, Santa Monica, Calif.: RAND Corporation, MG-403-AF, 2006. As of June 21, 2021: https://www.rand.org/pubs/monographs/MG403.html

Murray, Williamson, *Military Adaption in War: With Fear of Change*, Cambridge, UK: Cambridge University Press, 2014.

National Academies of Sciences, Engineering, and Medicine, *Quantum Computing: Progress and Prospects*, Washington, D.C.: National Academies Press, 2019.

National Intelligence Center, *Global Trends: Paradox of Progress*, January 2017. As of August 31, 2020: https://www.dni.gov/index.php/features/1685-nic-releases-global-trends-paradox-of-progress

Office of the Under Secretary of Defense (Comptroller), *European Deterrence Initiative: Department of Defense Budget, Fiscal Year (FY) 2020*, Washington, D.C., March 2019. As of June 21, 2021: https://comptroller.defense.gov/Portals/45/Documents/defbudget/fy2020/fy2020_EDI_JBook.pdf

Orme, John, "Deterrence Failures: A Second Look," *International Security*, Vol. 11, No. 4, Spring 1987, pp. 96–124.

Palmer, Jason, "Quantum Technology Is Beginning to Come into Its Own," *The Economist (Technology Quarterly: Here, There and Everywhere)*, 2017. As of July 5, 2021:

https://www.economist.com/news/essays/21717782-quantum-technology-beginning-come-its-own

Pecotic, Adrian, "Whoever Predicts the Future Correctly Will Win the AI Arms Race," *Foreign Policy*, March 5, 2019. As of July 9, 2021:
https://foreignpolicy.com/2019/03/05/whoever-predicts-the-future-correctly-will-win-the-ai-arms-race-russia-china-united-states-artificial-intelligence-defense/

Posen, Barry R., *Inadvertent Escalation: Conventional War and Nuclear Risks*, Ithaca, N.Y.: Cornell University Press, 1991.

Press, Daryl G., *Calculating Credibility: How Leaders Assess Military Threats*, Ithaca, N.Y.: Cornell University Press, 2005.

Rasler, Karen, and William R. Thompson, "Explaining Rivalry Escalation to War: Space, Position, and Contiguity in the Major Power Subsystem," *International Studies Quarterly*, Vol. 44, No. 3, September 2000, pp. 503–530.

Robinson, Michael, Kevin Jones, and Helge Janicke, "Cyber Warfare: Issues and Challenges," *Computers and Security*, Vol. 49, 2015, pp. 70–94.

Rosen, Stephen Peter, *Winning the Next War: Innovation and the Modern Military*, Ithaca, N.Y.: Cornell University Press, 1991.

Sayler, Kelley M., "Defense Primer: U.S. Policy on Lethal Autonomous Weapon Systems," Washington, D.C.: Congressional Research Service, IF11150, 2019. As of June 12, 2021:
https://fas.org/sgp/crs/natsec/IF11150.pdf

Sayler, Kelley M., *Artificial Intelligence and National Security*, Washington, D.C.: Congressional Research Service, November 21, 2019. As of March 5, 2020:
https://fas.org/sgp/crs/natsec/R45178.pdf

Sayler, Kelley M., and Laurie A. Harris, *Deep Fakes and National Security*, Washington, D.C.: Congressional Research Service, October 14, 2019, updated June 8, 2021. As of June 12, 2021:
https://crsreports.congress.gov/product/pdf/IF/IF11333

Scharre, Paul, *Army of None: Autonomous Weapons and the Future of War*, New York: W. W. Norton, 2019.

Schleher, D. Curtis, *Introduction to Electronic Warfare*, Dedham, Mass.: Artech House, 1986.

Schmitt, Michael N., ed, *Tallinn Manual 2.0 on the International Law Applicable to Cyber Operations*, 2nd ed., Cambridge, UK: Cambridge University Press, 2017. As of August 18, 2020:
https://ccdcoe.org/research/tallinn-manual/

Schneider, Jacquelyn, "The Capability/Vulnerability Paradox and Military Revolutions: Implications for Computing, Cyber, and the Onset of War," *Journal of Strategic Studies*, Vol. 42, No. 6, September 2019, pp. 841–863.

Schwab, Klaus, *The Fourth Industrial Revolution*, New York: Crown Business, 2016.

Sebbane, Yasmina Bestaoui, *Smart Autonomous Aircraft: Flight Control and Planning for UAV*, 1st ed., Boca Raton, Fla.: CRC Press, 2015.

Sechser, Todd S., Neil Narang, and Caitlin Talmadge, "Emerging Technologies and Intra-War Escalation Risks: Evidence from the Cold War, Implications for Today," *Journal of Strategic Studies*, Vol. 42, No. 6, 2019a, pp. 864–887.

Sechser, Todd S., Neil Narang, and Caitlin Talmadge, "Emerging Technologies and Strategic Stability in Peacetime, Crisis, and War," *Journal of Strategic Studies*, Vol. 42, No. 6, 2019b, pp. 727–735.

Shafiee, Elahe, Mohammad Reza Mosavi, and Maryam Moazedi, "Detection of Spoofing Attack Using Machine Learning Based on Multi-Layer Neural Network in Single-Frequency GPS Receivers," *Journal of Navigation*, Vol. 71, No. 1, 2018, pp. 169–188.

Shurkin, Michael, "Modern War for Romantics: Ferdinand Foch and the Principles of War," *War on the Rocks*, July 8, 2020. As of August 19, 2020:
https://warontherocks.com/2020/07/modern-war-for-romantics-ferdinand-foch-and-the-principles-of-war/

Tarraf, Danielle C., William Shelton, Edward Parker, Brien Alkire, Diana Gehlhaus, Justin Grana, Alexis Levedahl, Jasmin Léveillé, Jared Mondschein, James Ryseff, Ali Wyne, Dan Elinoff, Edward Geist, Benjamin N. Harris, Eric Hui, Cedric Kenney, Sydne Newberry, Chandler Sachs, Peter Schirmer, Danielle Schlang, Victoria Smith, Abbie Tingstad, Padmaja Vedula, and Kristin Warren, *The Department of Defense Posture for Artificial Intelligence: Assessment and Recommendations*, Santa Monica, Calif.: RAND Corporation, RR-4229-OSD, 2019. As of July 22, 2020:
https://www.rand.org/pubs/research_reports/RR4229.html

Thompson, Loren, "To Defeat Hypersonic Weapons, Pentagon Aims to Build Vast Space Sensor Layer," *Forbes*, February 4, 2020. As of June 21, 2021:
https://www.forbes.com/sites/lorenthompson/2020/02/04/space-sensor-layer-is-the-pentagons-next-tech-mega-project/?sh=17139552719d

United Kingdom Government, National Cyber Security Centre, "Whaling: How It Works, and What Your Organisation Can Do About It," webpage, October 6, 2016. As of July 21, 2020:
https://www.ncsc.gov.uk/guidance/whaling-how-it-works-and-what-your-organisation-can-do-about-it

U.S. Air Force Scientific Advisory Board, "Utility of Quantum Systems for the Air Force," study abstract, 2015. As of July 5, 2021:
https://www.scientificadvisoryboard.af.mil/Portals/73/Documents/Abstract/Abstract%20201 5/Final%20UQS%20Abstract%20(Approved%20for%20Public%20Release).pdf?ver=w3TG Fs59Sipu5fbKulUM7g%3d%3d

U.S. Department of Defense, *Militarily Critical Technologies*, Washington, D.C., September 19, 2001. As of June 21, 2021:
https://www.hsdl.org/?abstract&did=1109

U.S. Department of Defense, *Summary: Department of Defense Cyber Strategy 2018*, Washington, D.C., 2018a. As of June 21, 2021:
https://media.defense.gov/2018/Sep/18/2002041658/-1/- 1/1/CYBER_STRATEGY_SUMMARY_FINAL.PDF

U.S. Department of Defense, *Summary of the 2018 Department of Defense Artificial Intelligence Strategy: Harnessing AI to Advance Our Security and Prosperity*, Washington, D.C., 2018b. As of June 21, 2021:
https://media.defense.gov/2019/Feb/12/2002088963/-1/-1/1/SUMMARY-OF-DOD-AI- STRATEGY.PDF

U.S. Department of Defense, *Nuclear Posture Review*, Washington, D.C., February 2018c. As of June 21, 2021:
https://media.defense.gov/2018/Feb/02/2001872886/-1/-1/1/2018-NUCLEAR-POSTURE- REVIEW-FINAL-REPORT.PDF

U.S. Department of Defense, *United States Space Force*, Washington, D.C., February 2019a. As of June 21, 2021:
https://media.defense.gov/2019/Mar/01/2002095012/-1/-1/1/UNITED-STATES-SPACE- FORCE-STRATEGIC-OVERVIEW.PDF

U.S. Department of Defense, *Indo-Pacific Strategy Report: Preparedness, Partnerships, and Promoting a Networked Region*, Washington, D.C., June 1, 2019b. As of June 21, 2021:
https://media.defense.gov/2019/Jul/01/2002152311/-1/-1/1/DEPARTMENT-OF-DEFENSE- INDO-PACIFIC-STRATEGY-REPORT-2019.PDF

U.S. Department of Defense and Office of the Director of National Intelligence, *National Security Space Strategy: Unclassified Summary*, Washington, D.C., January 2011. As of June 21, 2021:
https://www.dni.gov/files/documents/Newsroom/Reports%20and%20Pubs/2011_nationalsec urityspacestrategy.pdf

U.S. Department of Homeland Security, and Office of the Director of National Intelligence, *Emerging Technology and National Security*, 2018 Analytic Exchange Program, July 26, 2018. As of August 31, 2020:
https://www.dhs.gov/sites/default/files/publications/2018_AEP_Emerging_Technology_and_National_Security.pdf

U.S. Government Accountability Office, *Report to Congressional Committees: National Security, Long-Range Emerging Threats Facing the United States as Identified by Federal Agencies*, Washington, D.C., GAP-19-204SP, December 2018. As of June 12, 2021:
https://www.gao.gov/assets/gao-19-204sp.pdf

U.S. Senate Committee on Armed Services, "Statement of Admiral Philip S. Davidson, U.S. Navy Commander, U.S. Indo-Pacific Command, Before the Senate Armed Services Committee on U.S. Indo-Pacific Command Posture," Washington, D.C., February 12, 2019. As of June 21, 2021:
https://www.armed-services.senate.gov/imo/media/doc/Davidson_02-12-19.pdf

U.S. Senate Committee on Armed Services, "Statement of General Terrence J. O'Shaughnessy, United States Air Force Commander, United States Northern Command and North American Aerospace Defense Command Before the Senate Armed Services Committee," Washington, D.C., February 13, 2021. As of June 21, 2021:
https://www.armed-services.senate.gov/imo/media/doc/OShaughnessy_02-26-19.pdf

Weisiger, Alex, and Keren Yarhi-Milo, "Revisiting Reputation: How Past Actions Matter in International Politics," *International Organization*, Vol. 69, No. 2, 2015, pp. 473–495.

The White House, *National Security Strategy of the United States of America*, Washington, D.C., December 2017. As of June 21, 2021:
https://trumpwhitehouse.archives.gov/wp-content/uploads/2017/12/NSS-Final-12-18-2017-0905.pdf

World Economic Forum, *Top 10 Emerging Technologies 2019*, Geneva, Switzerland, June 2019. As of June 21, 2021:
http://www3.weforum.org/docs/WEF_Top_10_Emerging_Technologies_2019_Report.pdf

Zabecki, David T., "Book Review: Military Adaption in War, by Williamson Murray," HistoryNet webpage, undated. As of July 9, 2021:
https://www.historynet.com/book-review-military-adaptation-in-war-by-williamson-murray.htm